Scientific and Technical Conference on Innovative Mining Technologies (IMTech 2019)

Part 2

IOP Conference Series: Materials Science and Engineering
Volume 679

Szczyrk, Poland
14 - 16 October 2019

Editors:

Marcin Lutynski
Dariusz Prostanski
Tomasz Suponik

ISBN: 978-1-7138-0931-9
ISSN: 1757-8981

Printed from e-media with permission by:

Curran Associates, Inc.
57 Morehouse Lane
Red Hook, NY 12571

Some format issues inherent in the e-media version may also appear in this print version.

Institute of Physics
Dirac House, Temple Back
Bristol BS1 6BE UK

Phone: 44 1 17 929 7481
Fax: 44 1 17 920 0979

techtracking@iop.org

Additional copies of this publication are available from:

Curran Associates, Inc.
57 Morehouse Lane
Red Hook, NY 12571 USA
Phone: 845-758-0400
Fax: 845-758-2633
Email: curran@proceedings.com
Web: www.proceedings.com

TABLE OF CONTENTS

IOP Conf. Series: Materials Science and Engineering **679** (2019) 011001 doi:10.1088/1757-899X/679/1/011001

INNOVATIVE MINING TECHNOLOGIES

IMTech 2019 Scientific and Technical Conference

March 25th – 27th, 2019

Szczyrk, Poland

The IMTech 2019 Scientific and Technical Conference on INNOVATIVE MINING TECHNOLOGIES was held in Szczyrk, Poland. It gathered over 100 scientists and researchers as well as a big group of specialists representing mining machinery producers and producers of minerals.

The IMTech 2019 Conference was oriented mainly onto innovative and environment friendly technologies and techniques for processing of minerals in the aspect of safety, quality and efficiency.

Five conference sessions contained interdisciplinary presentations on the latest achievements and implementations in the domain of minerals' beneficiation machines and equipment, a beneficiation and an exploitation of minerals, a review of technologies for a low-quality solid fuel gasification, dust control in hard coal processing plants, a remediation and a revitalization of industrial areas, a recovery of metals from e-scrap using gravitational, electrostatic and magnetic separations as well as underground mine workings as potential places for compressed air energy storage and a utilization of methane deposits from liquidated mines in the aspect of environmentally friendly energy technologies.

The conference was an interesting forum enabling an exchange of scientific and professional experience among the participants representing academia and industry. However, the conference series contains only some selected articles, giving an overview of the IMTech 2019 subject-matter.

Editors

Dr. Marcin Lutyński, DSc. Eng.   Dr. Dariusz Prostański, DSc. Eng.   Dr. Tomasz Suponik, DSc. Eng.

IOP Conf. Series: Materials Science and Engineering **679** (2019) 011002 doi:10.1088/1757-899X/679/1/011002

Peer review statement

All papers published in this volume of *IOP Conference Series: Materials Science and Engineering* have been peer reviewed through processes administered by the proceedings Editors. Reviews were conducted by expert referees to the professional and scientific standards expected of a proceedings journal published by IOP Publishing.

IOP Conf. Series: Materials Science and Engineering **679** (2019) 012001 doi:10.1088/1757-899X/679/1/012001

Historical background and development trends of minerals' processing – conclusions from KOMEKO–IMTech 2019 Conference

M Malec

KOMAG Institute of Mining Technology, Pszczyńska 37, 44-101 Gliwice, Poland

E-mail: mmalec@komag.eu

Abstract. The article describes historical background and a few development trends of minerals' processing on the base of the conclusions drawn during the KOMEKO-IMTech 2019 Conference organized by the KOMAG Institute of Mining Technology last March. The scope of the Conference covered an exchange of information and professional experience among representatives of science and industry technology. Some information about processing techniques and technologies in Soverign Poland are also presented in the article. Special attention is paid to innovative solutions of minerals' beneficiation machines and equipment, in particular developed at KOMAG for the hard coal mining industry.

1. Introduction

Each March the KOMAG Institute of Mining Technology organizes a scientific and technical conference on innovative and environment - friendly techniques and technologies of minerals' processing in the aspect of safety, quality and efficiency. This year the 20th KOMEKO – IMTech 2019 Conference was held in Szczyrk, Poland from 25th till 27th March 2019 and it gathered researchers, designers, producers and users of preparation plant machinery and equipment. Due to that an interesting and creative forum of specialists, ready to share their knowledge and professional experience, had a possibility to discuss the present issues as well as development trends of minerals' processing techniques and technologies. The subject-matter of the conference covered, among others, the following issues:

- Gasification technologies of solid fuels and ecofuels.
- State-of-the-art machines and equipment for processing of minerals.
- Modernization of minerals' preparation plants.
- Recovery of rare-earth elements from hard coal.
- A development of pulsation jigs, planetary mills and hydrocyclones.
- A recultivation and revitalization of post-industrial areas – coal recovery from waste dumps.

The KOMEKO-IMTech 2019 Conference was really special due to a small jubilee. In fact it was the 20th Conference from the KOMEKO cycle attended by numerous specialists, representing science and industry. It was sponsored by CARBOAUTOMATYKA, OPA CARBO MIFAMA, NORD DRIVE SYSTEMS and OMAG.

2. Development of hard-coal mining industry and of minerals' processing techniques and technologies in Sovereign Poland (1918-2018).

When Poland regained independence in 1918, it was very important to concentrate on developing its economy, as there were big needs as well as multi-year negligences caused by the war. The government activities were oriented onto Upper Silesia, where different minerals were deposited, mainly hard coal but also iron ore, zinc and lead. Before the first world war 250 thousand workers were employed in the industry and about 200 thousand of them worked in the mining and metallurgical branches of industry. Then there were 67 hard coal mines, 15 mines of zinc ore and 9 mines of iron ore. The hard coal production rate was 43.5 mln tons in 1918.

In the period between the wars the production rate of hard coal in Poland decreased significantly due to an economic crisis. During the second world war the hard coal production rate was only 20 mln tons. After the war, in 1946 the production rate of hard coal reached 51 mln tons. The production top rate was achieved in 1979, when it reached 201 mln tons. However, in 1990 the production rate of 70 coal-mines in operation was 151.3 mln tons.

In 2018 in total 18 coal-mines produced 63.4 mln tons of hard coal and there was a record import of coal in the range of 16.2 mln tons. A development of hard coal processing plants started during the period between the wars. It consisted in a modernization of machines and equipment in the existing preparation plants.

In practice, in all the coal mines, the run-of-mine was directed to preparation plants to be sorted manually. The modernization covered an installation of crushers, screens, washeries and air classifiers. It is important to mention that belt conveyors (which facilitated haulage activities significantly), were introduced then. An electric drive became common.

The production rate of the biggest sorting plants reached 600 t/h. Cascade washeries, installed in the Bytom and Hohenzollern Mines, were one of several innovative solutions of important character. In the thirties air classifiers were applied in the Bielszowice and Abwehr Mines. Their production rate was from 80 to 100 t/h in the case of grain sizes from 0.5 to 35 mm [1, 5]. A new design solutions concerned the smallest grains. In 1937-1938 the Dębieńsko Mine installed a flotation machine for the first time. It processed 75 thousand tons of concentrate within a year, using two batteries of the total production rate of 25 t/h. It should be noticed that the first washeries in the Upper Silesian mines were implemented twenty years after their introduction in England and in America. In 1946 the coal production rate in Poland was 51 mln tons and the processing capabilities were 19800 t/h. Over the period 1946-1964 some new preparation plans were constructed and several ones were modernized. In total 21 sorting plants, 19 washeries, 3 air classifiers and 6 flotation plants started their operational activity. In the same time 3 sorting plants, 4 washeries and 1 flotation plant were modernized. As the coal production rate was 130.7 mln tons in 1963, so 93% of the run-of-mine could be subject to preparation processes. It should be borne in mind that in 1946 the preparation plants could process 52.3% of the run-of-mine in the sorting plants and washeries [5]. The following years up to 1997 brought a significant progress in the scope of coal preparation plant development, as 53 new plants were built and 23 were modernized. The restructuring activities of the Polish mining industry, undertaken after 1989, had a strong impact on functioning of coal processing plants. Over the period 1989-1997 22 preparation plants were closed down due to a radical reduction of the coal production rate. In 1997 hard coal was mined in 57 mines and 66 preparation plants were in operation, where 66 grain washeries +20 (10) mm, 45 fines washeries 20 (10)÷0.5 mm and 22 coal flotation sections 0.5÷0 mm worked [1].

In 1997 the basic technological centres of preparation plants were equipped with machines and equipment presented in Table 1 [5].

Table 1. Equipment and machines operated in preparation plants in 1997 [5].

Item	Machines and equipment	Number of machines		
		in total	domestic	imported
1	Screens of preliminary classification	394	351	43
2	Beneficiation equipment for classes +0.5	417	371	46
3	Flotation machines	86	77	9
4	Dewatering machines	916	881	35
5	In total	1813	1680	133

At present in 18 hard coal mines there are 40 preparation plants in operation. Among them 33 plants beneficiate steam coal (9 of them beneficiate coal only in the classes above 20 (10) mm, 16 of them – in the classes above 0.1 mm and 8 of them – in the whole grain scope).

Coking coal is beneficiated in 7 coal processing plants within the whole scope of grain sizes. The machines and equipment, used in these processes, include: pulsating jigs, different types of screens, crushers, heavy – medium separators, Reichert spirals, heavy – medium cyclones, hydrocyclones, flotation machines, vibrating screens, and vibrating centrifuges, screen – sedimentation centrifuges, belt presses, filtration presses, pressure filters and disc vacuum filters. At present these technological nodes are automated and visualized.

3. Achievements of KOMAG Institute of Mining Technology in implementing innovative solutions in hard coal preparation plants.

The KOMAG Institute of Mining Technology specializes in designing and testing of machines and equipment for an exploitation as well as processing of minerals. It was established in 1950 and up till now underwent several organizational changes, enabling its adaptation to the market oriented economy. KOMAG has always been active and successful in implementing innovative machinery solutions which are operator and environment friendly [2]. Since 1950 over 1100 technical documentations of machines and equipment, used in mines of minerals in Poland and abroad, have been elaborated. An innovative character of these solutions is confirmed by more than 4400 patents and utility patents granted by the Polish Patent Authority.

Research projects in the domain of minerals' processing machinery and equipment have been realized since 1955. In this scope of activity KOMAG has extremely efficient solutions as regards DISA separators and pulsating jigs. Over the period 1955-2018 technical documentations of 320 separators and over 200 separators in pulsating water medium, i.e. pulsating jigs [4] have been designed at KOMAG. The present offer includes jigs of KOMAG type such as: OM – fines jigs for grain class

20-(0.5) mm, OS – medium size grain jigs for grain class 80 (50) – 0 (0.5) mm and OZ – grain jigs for grain class 120 – 20 mm. These jigs are operated not only in Poland but also in Argentina, Brazil, China, India, Romania and Vietnam. It is important to realize the fact that modified designs of jigs are also used for a production of aggregates. A list of jigs, designed over the years 1955-2018 and implemented in the Polish hard coal mines and in the aggregates production plants as well as installed abroad, is presented in Table 2.

Table 2. List of KOMAG type jigs designed over the year 1955 -2018 [4, 6].

Production period	Type of jig	Name of coal-mine
1955 - 1970	OBM 12, OBM 15, OBSZ15, OBZ 10, OBZ 12	Anna, Dębieńsko, Knurów, 1 – Maja, Nowy Wirek, Mysłowice, Rydułtowy Export to China, India and Vietnam
1971 - 1985	ODM 10, ODM 18, ODZ 15, OM 12, OM 12-2, OM-12-3, OM 12-3S, OM 12G3, OM 12 P3, OM 12 L3, OM 18P3, OM 18 L3, OM 24-3, OM 24D, OM 24 BC, OM 24 D3, OZ 18L, OZ 12, OZ 12L, OZ 12 P3, OS 36D3, OZ 36D3, OC8, OC10	Bogdanka, Borynia, Dębieńsko, Gliwice, Halemba, Knurów, Krupiński, Jankowice, Makoszowy, Marcel, Pniówek, Dymitrow, Rydułtowy, Sośnica, Staszic, Wujek, Szczygłowice, Wawel, Zabrze, Zofiówka, Moszczenica Export to Brazil, India and Romania
1986 - 2014	OM 8L2e, OM 8L2E, OM10L2E, OM 15 B3E, OM 12P3E, OMPE-3x6,5, OM 18L3E, OM 18P3E, OM 20 P3E, OM 20L3E, OM 24 P4E, OM 24L4E, OM 24D3E, OM 30-3E, OM 18 3x8, OM30 D3E, OS30 D3E, OM 30, OZ 18 L3E, KOD jig	Andaluzja, Anna, Barbara – Chorzów, Bogdanka, Borynia, Budryk, Dębieńsko, Halemba, Jastrzębie, Jasmos, Knurów, Krupiński, Marcel, Pniówek, Rozbark, Rymer, Rydułtowy – Anna, Sośnica, Staszic, Szczygłowice, Wawel, Wujek, Zofiówka Export to Czech Republic and India Budryk – for waste removal from run-of-mine
2003 - 2018	Pulsating classifiers: K - 100, K – 150, K – 50, K – 80, K – 101, K – 102, K - 151	KSM in Borzęcin, PPMDKruszbet in Suwałki, PRInż Surowce in Januszkowice, PUHM “M” in Kędzierzyn – Koźle, Żwirownia Bierawa, PRESTO – Emil Potręć, Rokitno, Rent – Pol – Przezchlebie
2014 - 2019	OM 30, OS18L, OS18P, OM15L, OM 15P, OS 4, OM 20, OM18L, OS 18P, OM 24, OM 15L, OM 15P	Sośnica, Krupiński, Pniówek, Budryk, Zofiówka, ZG „Eko-Plus”, ZG Sobieski

Research and development projects, realized at the KOMAG Institute within recent years were also oriented onto hydrocyclones, screens and crushers. They concerned not only technical and technological aspects, but their operational safety in particular. One of the hazards, experienced by the personnel of preparation plants, is dust. Four different types of dust control equipment are used:

- UO type using a wet dust control method, where water is dispersed by means of fixed and rotational nozzles; the dust control efficiency reaches 99.7%,
- DCU type, where a stream of dust laden air contacts with water curtain generated by a rotor, driven by a motor, which acts as a fan forcing the air under control,

IOP Conf. Series: Materials Science and Engineering **679** (2019) 012001 doi:10.1088/1757-899X/679/1/012001

- LCDU type, where a mechanism of wet dust control connected with labyrinth air flow, is used. Due to frequent and violent direction changes of dust laden air particles, they loose their momentum and are easy to be separated on the water curtain. This dust collector can operate together with axial-flow as well as centrifugal fans, installed in front of or behind the dust collector which enables an operation in sucking or combined ventilation systems.
- DRU type in which a traditional mechanism of water spraying, connected with a centrifugal separation, achieved due to use of two guide vanes, gives a high efficiency at small resistance to flow.

Another efficient dust control method, developed at KOMAG, consists in an application of spraying equipment in a form of a battery with spraying nozzles which are installed in the areas where dust is generated on transportation routes of the run-of-mine. These installations are supplied with water and compressed air. One of the solutions is the PASAT mist system installed on discharge points of conveyors and on crushers. This system was implemented in the Bolesław Śmiały and Mysłowice -Wesoła Coal Mines. Only recently the NEPTUN system, consisting of seven independent installations, built on five work-places, was implemented in the Bolesław Śmiały Coal-Mine.

4. Examples of modernization activities in hard coal preparation plants.

4.1. Modernization of the Pniówek Preparation Plant

The modernized Pniówek Preparation Plant started its operation in 2017. The beneficiation node consists of 8 fines jigs of OM 15L and OM 15P types (20÷0 mm) shown in Figure 1 and 8 bucket conveyors of B-1000 type shown in Figure 2.

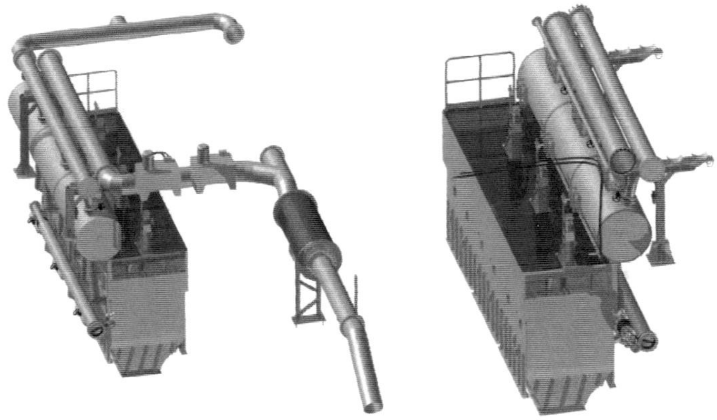

Figure 1. OM 15 Fines jig [source: KOMAG].

Figure 2. B-1000 Bucket conveyor [source: KOMAG].

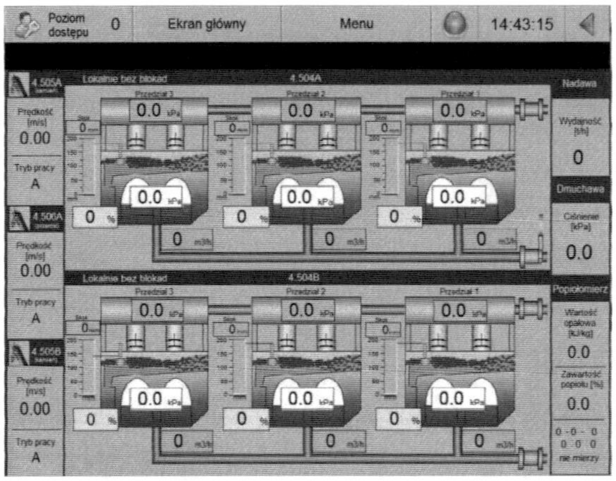

Figure 3. Jig node control system of KOGASTER type [source: KOMAG].

The jig node, shown in Figure 3 is controlled by the KOGASTER system developed at the KOMAG Institute.

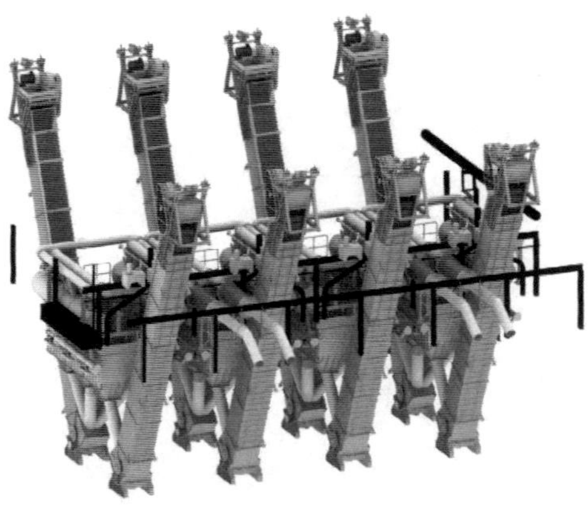

Figure 4. Modernized preparation plant at Pniówek Coal Mine.

It is important to emphasize the fact that in the case of the Pniówek Mine four two-trough jigs (Figure 4) were replaced by eight one-trough jigs of OM 15 type which facilitated the control possibilities of the beneficiation process.

4.2. Modernization of the Budryk Preparation Plant

The modernized Budryk Preparation Plant started its operation in 2018. The beneficiation node consists of 6 medium-size grain jigs of OS 18 (L and P) (70—2 mm) type, shown in Figure 5, 2 fines jigs OM 20 (12-0 mm) type, shown in Figure 6, and 14 belt conveyors of B-1000 type (Figure 7).

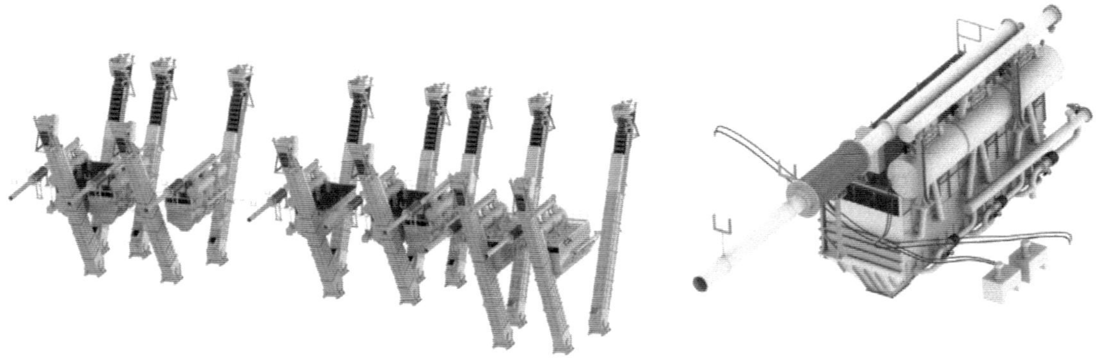

Figure 5. OS 18 medium-size grain jig [source: KOMAG].

Figure 6. OM 20 fines jig [source: KOMAG].

Figure 7. B-100 Bucket conveyor [source: KOMAG].

The jig node is controlled by the KOGASTER system developed at KOMAG. It is shown in Figure 3.

Due to a modernization of the Budryk Preparation Plant it was possible to beneficiate coal of two types: 34 and 35, using two independent systems.

4.3. Modernization of the Sośnica Preparation Plant

The KOGASTER SSWO system [3, 7] at the Sośnica Preparation Plant controls an operation of two OM 30 fines jigs for a three-product beneficiation of the run-of-mine in the grain class 20-0 mm. Three B-1000 bucket conveyors of an increased bucket capacity up to 120 cm^3 are installed. An implementation of new machines and equipment enabled to increase a supervision of the beneficiation process and to improve an efficiency of the run-of-mine separation in the jig node.

The structure of the jig node control system as well as the measured parameters [3] are shown in Figure 8.

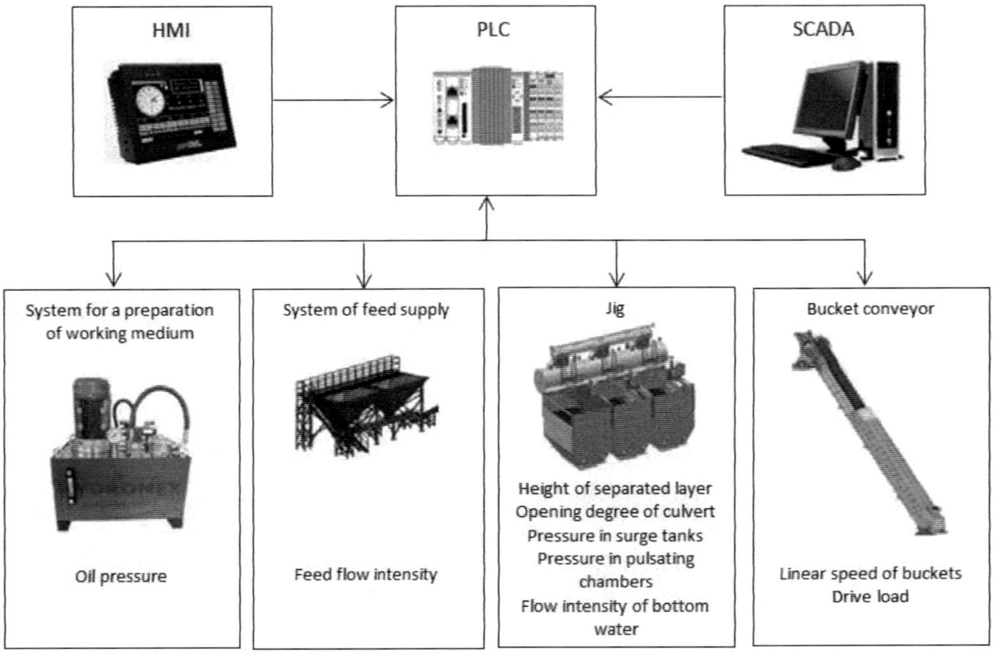

Figure 8. Structure of the jig node control system [3, 7, 8].

An automatic control of the hard coal beneficiation process in jigs is realized by:

- collecting data on the object condition,
- data processing enabling to take decisions concerning the process control,
- adjusting and taking corrective measures.

The control system, installed in the Sośnica Mine, is enables a control of the speed of the bucket conveyors, adapting their capacity to current loading [7, 9].

An installation of the KOAGSTER SSWO control system, elaborated at the KOMAG Institute, contributed to an increase of the Sośnica Preparation Plant efficiency and reliability.

4.4. Modernization of the Sobieski Preparation Plant.

In 2019 the jigs for the Sobieski Preparation Plant will be commissioned. The modernized beneficiation node consists of two fine coal jigs OM15L and OM15P for the grain class of 30-2 mm at the maximum capacity of 320 t/h each. They will replace two compartment Allmineral jigs. The

OM15 jigs will beneficiate three products. Modern WZL1 2.8x6.0 screening machines for separating the material below 2 mm are shown in Figure 9.

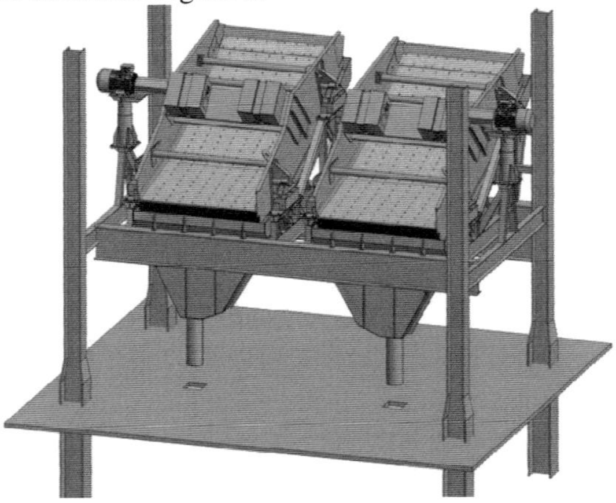

Figure 9. WZL1 2.8x6.0 screening machines installed at the Sobieski Preparation Plant [source: KOMAG].

5. Conclusions

- The technical and technological changes implemented in minerals', mainly hard coal preparation plants over the period of a hundred years of the Poland's sovereignty, contributed to a rapid development of the mining industry in Poland.
- A modernization of the existing and a construction of new preparation plants were realized with a close collaboration with the researchers from the KOMAG Institute of Mining Technology.
- Within the period of nearly 70 years of activity in the domain of designing and testing of mining as well as preparation plant machinery and equipment KOMAG researchers elaborated over 1100 technical documentations being the basis for manufacturing the machinery used in Poland and abroad (Argentina, Brazil, China, India, Romania and Vietnam).
- An innovative character of the KOMAG technical and technological solutions is confirmed by more than 4400 patents and utility patents granted by the Polish Patent Authority.
- Research and development projects, realized at the KOMAG Institute within recent years, concerned not only technical and technological aspects, but also widely understood operational safety, in particular dust control.
- As far as modernization activities in hard coal preparation plants are concerned three examples are given: Sobieski, Budryk and Sośnica Coal Mines, but special attention is paid to the KOGASTER jig node control system developed at KOMAG.

References

[1] Blaschke W and Lutyński A Przeróbka mechaniczna w kopalniach węgla kamiennego niepodległej Polski, Materiały konferencyjne KOMEKO 2019, Szczyrk, 25-27 marca 2019
[2] Kłeczek Z, Kozieł A and Malec M 60 lat działań Instytutu Techniki Górniczej KOMAG na rzecz rozwoju przemysłu maszyn górniczych. Maszyny Górnicze 3-4. 2010.
[3] Król J, Krzak Ł, Jendrysik S, Stankiewicz K and Woszczyński M Materiały konferencyjne KOMEKO 2019, Szczyrk, 25-27 marca 2019
[4] Matusiak P and Kowol D Rozwój osadzarek pulsacyjnych typu KOMAG. *Maszyny Górnicze* 2018 no **2**
[5] Stan przeróbki mechanicznej węgla kamiennego w Polsce. Opracowanie Agencji Restrukturyzacji Górnictwa Węgla Kamiennego S.A., Katowice, maj 1977

[6] Matusiak P and Kowol D State-of-the-art pulsating jigs of KOMAG type. Materiały na konferencję: IMTech 2019, Innovative Mining Technologies, Scientific and Technical Conference, Szczyrk, Poland, 25-27 March 2019 s. 1-10, (IOP Conference Series: Materials Science and Engineering 2019 vol. 545), ISSN 1757-899X.

[7] Jendrysik S and Kost G.: Control of bucket conveyor's output. Mechatronics 2017 - Ideas for Industrial Applications, Editors: Jerzy Świder, Sławomir Kciuk, Maciej Trojnacki, Advances in Intelligent Systems and Computing 934, Springer Nature Switzerland AG 2019 s. 192-200, ISBN 978-3-030-15857-6; ISSN 2194-5365.

[8] Rogala-Rojek J, Stankiewicz K and Jendrysik S SCADA class software of the KOGA control system of jig beneficiation node. Materiały na konferencję: IMTech 2019, Innovative Mining Technologies, Scientific and Technical Conference, Szczyrk, Poland, 25-27 March 2019 s. 1-12, (IOP Conference Series: Materials Science and Engineering 2019 vol. 545), ISSN 1757-899X.

[9] Kowol D and Matusiak P Improving the quality of hard coal products using the state of the art KOMAG solutions in a pulsating jig nod. Materiały na konferencję: MEC 2019, Mineral Engineering Conference, Kocierz, 16-19 September 2019, ISSN 1757-899X.

IOP Conf. Series: Materials Science and Engineering **679** (2019) 012002 doi:10.1088/1757-899X/679/1/012002

Simulation tests of new solution of the longwall shearer haulage system

K Kotwica[1]*, G Stopka[1] and P Gospodarczyk[1]

[1]AGH University of Science and Technology, Faculty of Mechanical Engineering and Robotics, Al. Mickiewicza 30, 30-059 Cracow, Poland

*E-mail: kotwica@ahg.edu.pl

Abstract. The article presents the idea of a new solution for KOMTRACK longwall shearer haulage system. Based on the analysis of test results of an earlier version of the FLEXTRACK haulage system, a new version of the KOMTRACK system was developed. Using the ANSYS LS-Dyna program, simulations of the longwall shearer's drive wheel mating with the selected elements of the toothed haulage segment were performed. The purpose of the analyzes was identification of contact stress value of the elements of the feed system as a function of geometrical parameters of various solutions of the toothed bar. The result of the simulation was the selection of the rack solution for further field and industrial tests.

1. Introduction

In polish underground coal mines currently hard coal is excavated in longwall system, using mainly shearers. The EICOTRACK system is a widely used longwall shearer haulage system. In this system, the shearer is moved by the interaction of the drive teeth wheels with the rack (ladder) located horizontally in the face conveyor gate. The segments of the racks (ladders) are fixed to the bracket supports with bolts in a rigid manner. The vertically positioned drive teeth wheel of the shearer tractor engages with its teeth with horizontally positioned pins of the ladder, forcing the change of the location of the shearer. The leading part of the shearer moves along the ladder in a sliding way, providing constant conditions for the cooperation of the meshing between the drive teeth wheel and the ladder [1, 2].

This results in accelerated wear of the drive wheel teeth and leads to frequent changes of drive wheels during the operation of the longwall work (3-4 times depending on the wall run and mining and geological conditions). This process generates additional costs for shearers producers and users and also creates the need to work of shearers servicemen in dangerous conditions. Downtime in mining, forced by maintenance work, negatively affects the economic result. The stress analysis carried out by ITG KOMAG shows that in the course of operation, the yield point of the material from which the wheel is made is exceeded. This is the result of a rigid attachment of ladders, which causes that in the case of transverse or vertical bending of the face conveyor, there is a so-called cutting the tooth of the drive wheel and disturbing of the pitch, i.e. the distance between the neighbouring teeth. The described problem is intensified by the increasing power of shearer tractors, which has now reached the value of 120 kW, causing a significant exceeding of the permissible surface pressure between the cooperating surfaces of the drive wheels and the ladder [3].

IOP Conf. Series: Materials Science and Engineering **679** (2019) 012002 doi:10.1088/1757-899X/679/1/012002

As a part of the project entitled "Innovative advance system of mining machine increasing the effectiveness of output and work safety in longwall systems", realized by the KOMAG Institute of Mining Technology (leader), Faculty of Mechanical Engineering and Robotics at the AGH University of Science and Technology, Foundry Research Institute and Specodlew Enterprise of Foundry Innovation Ltd., new solution of haulage system for longwall shearers was developed. It was called FLEXTRACK [1, 2, 3].

This solution eliminates mentioned above, unwanted collaboration conditions of the described kinematic pair. It is an innovative longwall shearer haulage system with flexible toothed segments. The toothed segments installed in guides, replacing the Eicotrack toothed bar (ladders), enable a more advantageous collaboration of the shearer driving wheel, in particular in the case of the curved conveyor line – it allows turning of flexible toothed segments both in the vertical as well as the horizontal planes. The operation principle of new haulage system FLEXTRACK and currently used system EICOTRACK is presented on the model in Figure 1 [4].

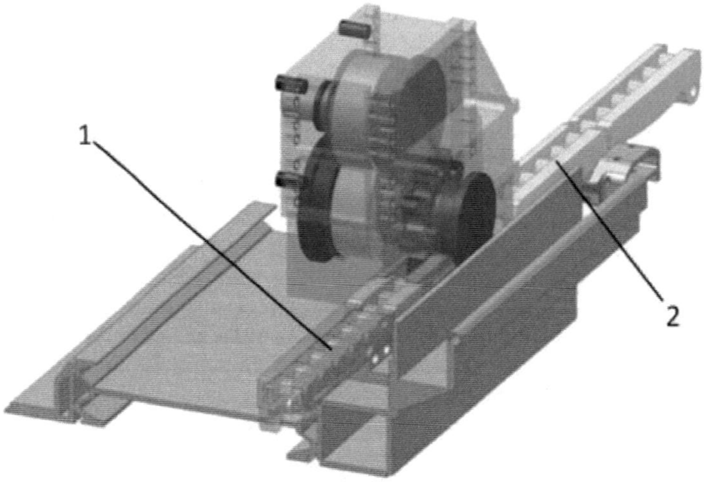

Figure 1. Model of collaboration of tractor drive toothed wheel with:
1 - system FLEXTRACK, 2 - system EICOTRACK [4].

The concepts of eight rack segment variants and five guides concepts were developed, which were subjected to strength, material and technological analysis.

In the rack segment variants (instead of the convex tooth profile used so far), a profile with a concave cylindrical lateral surface was proposed, whose radius was 5% larger than the largest radius of the driving teeth wheel. The results of the analysis allowed the selection and manufacturing of a prototype set of the most advantageous version of the rack segment and guide [4]. It has been subjected to field comparative tests with the ladders of the EICOTRACK system, using a longwall shearer and route elements of a longwall face conveyor. During the tests, for a comparable haulage lenght of the shearer, the qualitative and quantitative wear of drive wheels, ladders of the EICOTRACK system and rack elements of the FLEKTRACK system were measured, as well as the temperature of these elements. Selected results are presented below.

2. Selected field test results of the FLEXTRACK haulage system

The view of the test stand and the elements of the EICOTRACK and FLEXTRACK haulage system selected for testing are shown in Figures 2 and 3 [5]. The test stand allowed to simulate the bending of the longwall face conveyor route in a vertical and horizontal plane. In the case of FLEXTRACK haulage system shearer moved a total of about 7 km. In the case of EICOTRACK haulage system shearer moved about 1 km less. During tests with the EICOTRACK system, a significantly higher noise level was recorded, as well as greater vibration of the shearer and its load. In addition, several shearer failures occurred during the tests. In the case of wear and temperature measurements of the EICOTRACK and FLEXTRACK system elements, selected results are shown in Figures 4, 5, 6 and 7.

Figure 2. View of test stand for EICOTRACK and FLEXTRACK haulage system testing [5].

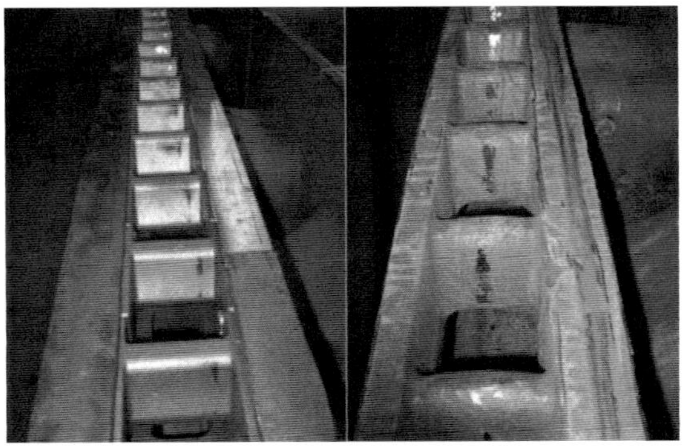

Figure 3. View of the elements of the EICOTRACK (right) and FLEXTRACK (left) haulage system selected for testing [5].

Figure 4. View of the wear of elements for FLEXTRACK haulage system after the test: on top – drive wheel, bottom – rack segments [5].

Figure 5. View of the wear of elements for EICOTRACK haulage system after the test: on top – drive wheel, bottom – ladder [5].

Based on the analysis of the obtained results, it can be stated that for the FLEXTRACK haulage system the shearer adapted better during the passage on the bent sections of the longwall face conveyor. The temperature of the system components, both racks or ladders, and drive wheels, was lower in the case of the FLEXTRACK haulage system. However, the quantitative (volumetric) wear of both cooperating elements was for the FLEXTRACK haulage system by several percent higher. However, the form of this wear was very even [5]. These elements, primarily the drive wheels, were

IOP Conf. Series: Materials Science and Engineering **679** (2019) 012002 doi:10.1088/1757-899X/679/1/012002

mainly wearing off. In the case of the EICOTRACK haulage system on the drive wheels, there were signs of chipping.

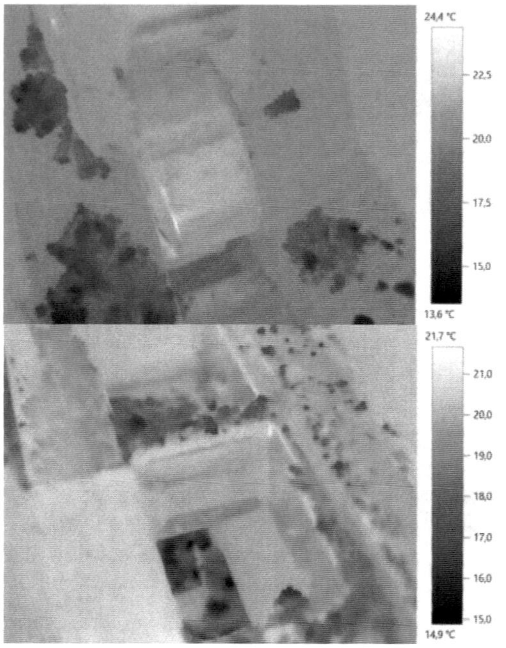

Figure 6. View of the temperature distribution of elements for FLEXTRACK haulage system after moving around 50 m: on top – drive wheel, bottom – rack segments [5].

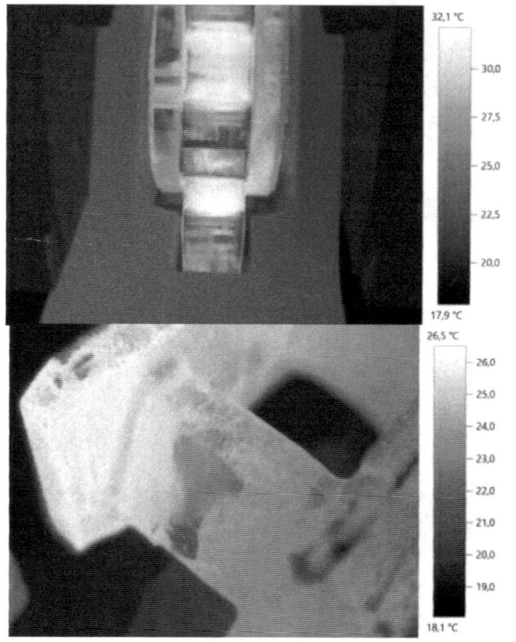

Figure 7. View of the temperature distribution of elements for EICOTRACK haulage system after moving around 50 m: on top – drive wheel, bottom – ladder [5].

Therefore, the new haulage system has proved its usefulness for work in difficult mining and geological conditions, when moved along a bent face conveyor route. However, the results of the study led to the conclusion that the proposed rack profile with a concave cylindrical lateral surface may be the reason for the increased wear of the elements of the FLEXTRACK haulage system. Therefore, as part of the new project "Innovative haulage system for highly productive longwall complexes" it was proposed to continue the topic and testing of the newly designed shearer haulage system called KOMTRACK .

A distinguishing feature of this system is also its flexibility enabling individual toothed segments (racks) to adapt to longitudinal and transverse curvatures of the longwall face conveyor route, which guarantees a correct cooperation with the drive wheel. The present knowledge, based on the results of the FLEXTRACK project, indicates that a flexibility of the racks, in relation to the rolling driving wheel, decreases a level of stresses generated during a longwall shearer displacement.

The following assumptions were accepted for reaching the project objective:

- the maximal force carried by a tooth of the toothed bar segment is 500 kN,
- the angle of rack turn in relation to each other is at least +/- 0.3 deg,
- values of stresses in each collaboration point of the frictional pair are below the material yield point,
- the shearer haulage system pitch is 151 mm.

Considering the above, it began to develop a construction form of a toothed segment (rack) adapted for a collaboration with the driving toothed wheel of a longwall shearer haulage unit.

However, it was proposed to change the shape of the rack. Instead of the concave surface, several versions of the convex or straight surface were selected.

IOP Conf. Series: Materials Science and Engineering **679** (2019) 012002 doi:10.1088/1757-899X/679/1/012002

To determine which version is the most favorable for further research, simulation tests were performed using the Ansys LS-Dyna program. The course and results of these tests are presented below.

3. Testing metodology

The physical model used in simulation was presented on Figure 8. The main assumption was to fix the drive wheel and to impose *constant angular velocity* ω = 0,5 1/s. The feed system consists of the two racks. The constrains were imposed on the bottom surface of the racks. The configuration of the degrees of freedom is illustrated in figure below. As it is illustrated, the haulage system has capability to free displacement only in the X direction. On the front surface of the rack a horizontal force of 335 kN was applied. In this way resistance to motion of a longwall coal shearer were simulated. Between the elements of the haulage system the surface contact were assumed so as to ensure free interaction between the haulage system elements. The simulation tests carried out for linear material properties. Material properties used in the simulation were presented in Table 1. The three types of the rack design used in the simulation tests are shown in Figure 9. They have different teeth profiles - flat, trapezoid of 20° wall inclination, convex with two curvature radius and convex with flat intermediate area inclined at 20°.

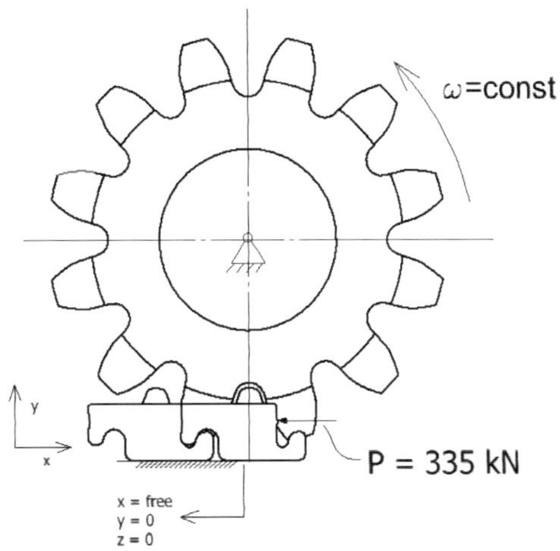

Figure 8. Schematic arrangement of testing model.

Table 1. Material properties.

Element	Young's modulus	Poisson ration	Yield strength
Drive wheel	207000 MPa	0.3	1080 MPa
Rack	210896 MPa	0.286	-

4. Results of simulation tests

Simulation tests has been conducted in Ansys LS-Dyna program. The results of the simulation tests was presented on Figures 10÷12. To simulate dynamic interaction between elements of the feed system the implicit algorithm was used. The aim of simulation tests was to identify the maximum equivalent stress, that occurs at the contact between the drive wheel and the racks, as a function of geometrical parameters of the rack construction. On figures below the stress concentration on the rack construction are shown. The location of stress concentration is changing because of dynamic operating between the drive wheel and the rack construction. In order to better show the stress distribution the equivalent stress above 1000 MPa has been marked in pink color. For the first type of the rack the maximum equivalent stress appears on the top edge of the second rack and the level of stress

IOP Conf. Series: Materials Science and Engineering **679** (2019) 012002 doi:10.1088/1757-899X/679/1/012002

significantly exceeded the value of 1000 MPa. The tooth profile modification for type V2 and V3 allowed to reduce level of the equivalent stress, as shown in Figures 11 and 12.

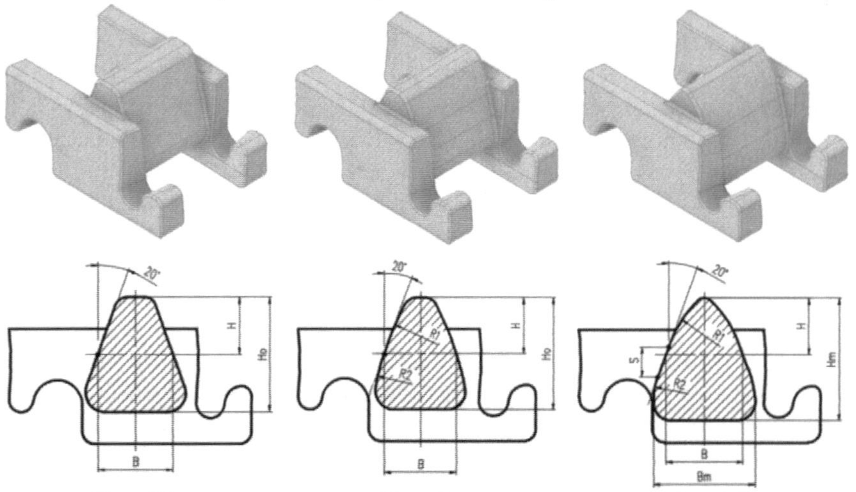

Type V1 **Type V2** **Type V3**

Figure 9. Types of rack segments of different teeth profiles used in simulation: Type V1 - flat, trapezoid of 20° wall inclination, Type V2 - convex with two curvature radius, Type V3 - convex with flat intermediate area inclined at 20°.

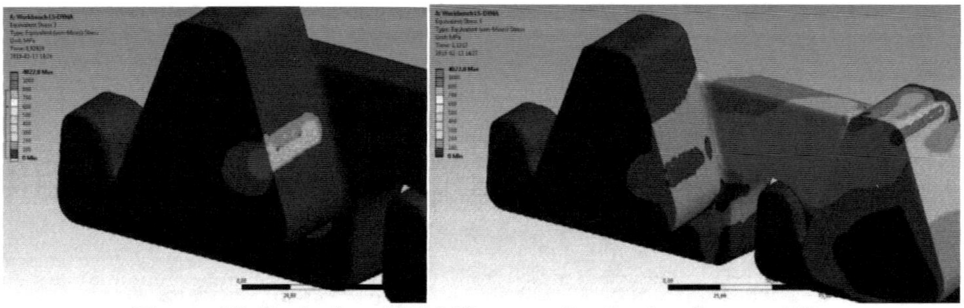

Figure 10. Equivalent (H-M-H) stress distribution for type V1.

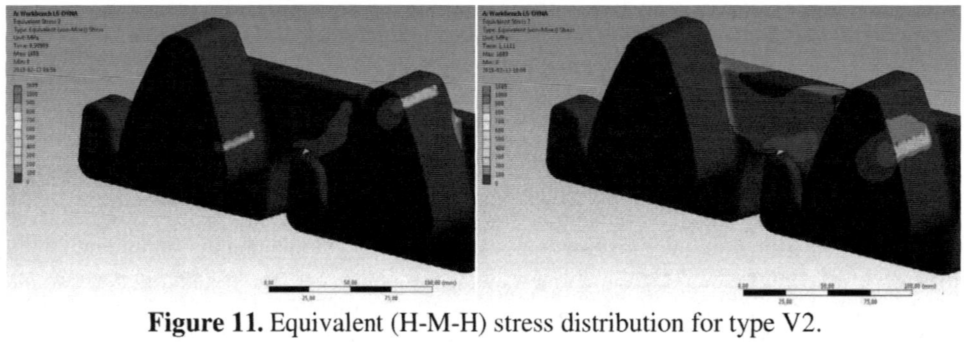

Figure 11. Equivalent (H-M-H) stress distribution for type V2.

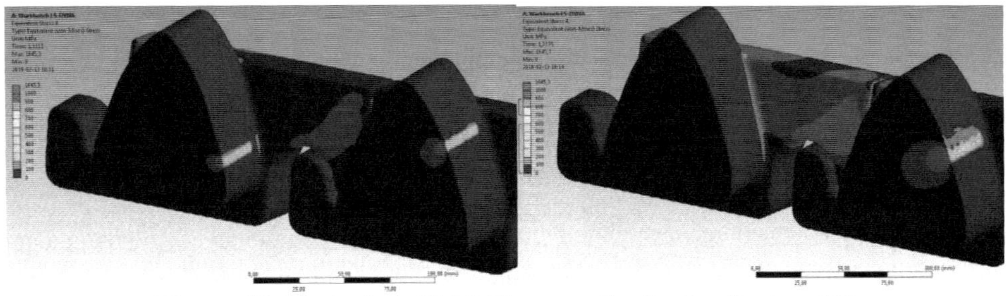

Figure 12. Equivalent (H-M-H) stress distribution for type V3.

Analyzing the stress distribution on the flank of the racks it can be concluded that the level of equivalent stress for type V3 is lower than in the case of type V2. The maximum equivalent stress on the flank of the rack V2 reached 1103 MPa. In the case of type V3 the equivalent stress reached 988 MPa.

5. Summary

The conducted research allowed to compare the impact of the tooth profile on the equivalent stress in the racks construction. In the field of dynamic computation, the three type of racks have been verified. The purpose of the study was to identify the stress concentration, that occurs at the contact between the drive wheel and the rack. The final results of the research was the recommendation of the selected rack design to the future laboratory test. The rack segment type V3 with convex and flat intermediate area inclined at 20° has been recommended. On the basis of designed CAD model the first prototype of the rack segment was manufactured. It is shown in Figure 13. This solution will be checked in the future during field and industrial tests. It is assumed that the working conditions of the examined system will be as close as possible to the conditions prevailing in the underground coal mines.

Figure 13. View of the first prototype of the rack segment type V3 with convex and flat intermediate area.

Acknowledgements

The article was created as part of project "The innovative haulage system for highly productive longwall complexes", according to the Contract No. POIR.04.01.04-00-0068/17 with the National Center for Research and Development, co-financed by the European Union from the European Regional Development Fund under the Intelligent Development Program.

Fundusze Europejskie Inteligentny Rozwój | Rzeczpospolita Polska | Narodowe Centrum Badań i Rozwoju | Unia Europejska Europejski Fundusz Rozwoju Regionalnego

References

[1] Zachura A et al. 2014 Projekt Program Badań Stosowanych - Innowacyjny system posuwu maszyny wydobywczej podnoszący efektywność wydobycia oraz bezpieczeństwo pracy w kompleksach ścianowych - Zadanie 5. Kryterialny wybór z pola możliwych rozwiązań oraz opracowanie postaci konstrukcyjnej elementów systemu posuwu maszyny wydobywczej. Instytut Techniki Górniczej KOMAG, Gliwice (not published)

[2] Zachura A and Żuczek R 2014 System posuwu Flextrack zwiększający trwałość i niezawodność napędu kombajnu ścianowego.[w:] Monografia: KOMTECH 2014, Innowacyjne techniki i technologie dla górnictwa. Bezpieczeństwo - Efektywność - Niezawodność, Instytut Techniki Górniczej KOMAG, Gliwice s. 151-161.

[3] Zachura A 2014 Quasi-statyczna analiza współpracy koła napędowego i zębatki w systemie posuwu kombajnu ścianowego *Mechanik* R. 87, nr **10**, s. 842—848

[4] Zachura A et al. 2015. Flextrack - innowacyjny system posuwu kombajnu ścianowego, cz.1 i 2. Nowoczesne metody eksploatacji węgla i skał zwięzłych: Monografia. Kraków: Wydawnictwa AGH, s. 185-204

[5] Kotwica K, Stopka G, Gospodarczyk P et al. 2016 Projekt Program Badań Stosowanych - Innowacyjny system posuwu maszyny wydobywczej podnoszący efektywność wydobycia oraz bezpieczeństwo pracy w kompleksach ścianowych - Zadanie 13. Badania stanowiskowe odlewów w warunkach laboratoryjnych. Akademia Górniczo-Hutnicza w Krakowie (not published)

IOP Conf. Series: Materials Science and Engineering **679** (2019) 012003 doi:10.1088/1757-899X/679/1/012003

Development of technology for withdrawal of the powered roof support from a row and its relocation from the liquidated longwall system

K Mazurek[1]*, M Szyguła[1] and K Turczyński[1]

[1]KOMAG Institute of Mining Technology, Division of Powered Roof Supports, Pszczyńska 37, 44-101 Gliwice, Poland

*E-mail: kmazurek@komag.eu

Abstract. Directions in improvement of efficiency and safety of the used systems for relocation of powered roof supports from the liquidated longwall was indicated basing on a review of the current systems. Suggestion for the machines and devices used for mechanization of a powered roof support relocation from a row and its transportation in cross-cut entry is presented. The innovations consist in combination of two functions in one machine used for roof support reinstallation and its rotation as well as its safe and fast relocation to the main transportation roadway.

1. Introduction

Necessity of reducing the number of operating mining fronts due to economic reasons makes the process of re-equipping of liquidated longwall panels and equipping new ones very important for effective mine operation. Such a process has to be carried out in a short time in an effective and safe way. Relocation of powered roof support is one of difficult technological operation due to high weight of the roof support and its big size, which is almost the same as the size of transportation routes. Additionally, the local mining-and-geological conditions such as convergence of the workings, roof conditions or increased methane concentration make the process more difficult [1]. Complexity of the roof support withdrawal from a row and its transportation in a cross-cut entry, specificity of local mining-and-geological conditions as well as different supporting technical equipment cause that the process is realized in each mining plant in a different way.

Experience shows that transporting the roof support from the liquidated longwall to a new longwall location depends on the proper preparation of the roof support for loading and the selection of a suitable transportation system. Preparation of the roof support for loading consists in pulling it out (withdrawal) from a row in the liquidated mining longwall, rotating and positioning along the created transportation channel of the cross-cut entry and dragging it to the loading place in the transportation system. Due to the shape of the cross-cut entry and its equipping, proximity of the roof fall zone and big mass of the roof support, these activities are time consuming, complicated and dangerous.

Current technologies of the roof support withdrawal form a row and its transportation in the cross-cut entry are analysed in the aspect of their effectiveness and safety as well as the concept of the innovative system for powered roof support withdrawal and its transportation in the cross-cut entry is presented.

2. Analysis of current technologies for powered roof support withdrawal from a line and its transportation in the cross-cut entry

To assess state of the art of the technology for withdrawal and transportation of powered roof support in the cross-cut entry, the following stages of the process were distinguished:

- withdrawal of the powered roof support from a row to the cross-cut entry,
- transportation of the powered roof support from the cross-cut entry to the main transportation roadway.

Using the practical experience of PGG S.A., JSW S.A mines as well as the independent mines the following variants are used for withdrawal of the roof support from a row:

With use of additional wooden sprag

Opposite the powered roof support, the wooden sprag, friction or hydraulic prop is installed at cross-cut entry front, which is additionally tied by a chain to the steel canopy located in a transportation area. The end of the beam of the roof support's advancing system is connected with the sprag and after the roof support yielding it is dragged to the cross-cut entry with use of the advancing ram. Then a winch rope is attached to the beam of advancing ram and by using it, the roof support rotates in such a way that it is positioned parallelly to the cross-cut entry front. At the place of withdrawn roof support, the crib boxes are constructed (Figure 1).

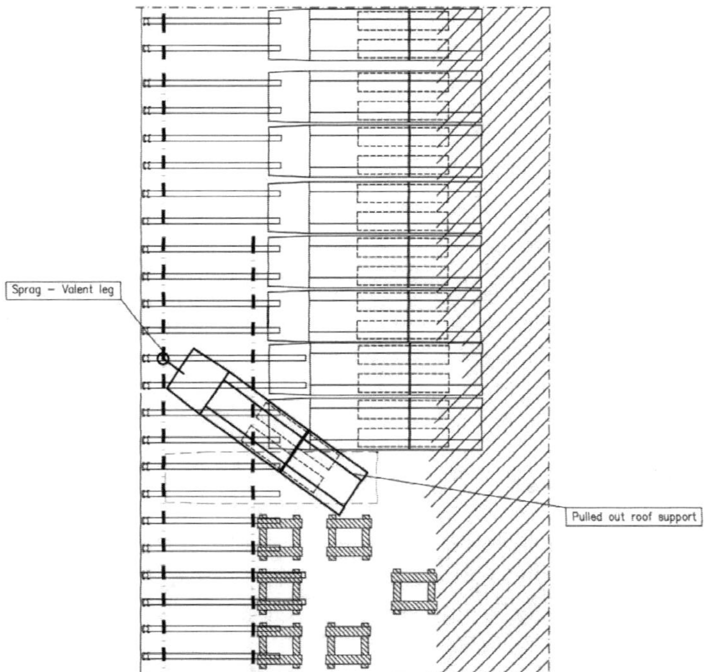

Figure 1. Powered roof support withdrawal with use of a wooden sprag.

With use of insurance roof support

First roof support withdrawn from a row and positioned in the cross-cut entry parallelly to its front plays a role of insurance roof support called sometimes closing support [2]. Its task is to protect the roof in the area of withdrawal of next roof supports as well as to play a role of a sprag during roof support withdrawal from a row. Withdrawn roof support, yielded to the height, which enables its movement under the insurance roof support, is dragged to the cross-cut entry using its advancing ram, and then it is positioned parallelly to the cross-cut entry front using the winch (Figure 2).

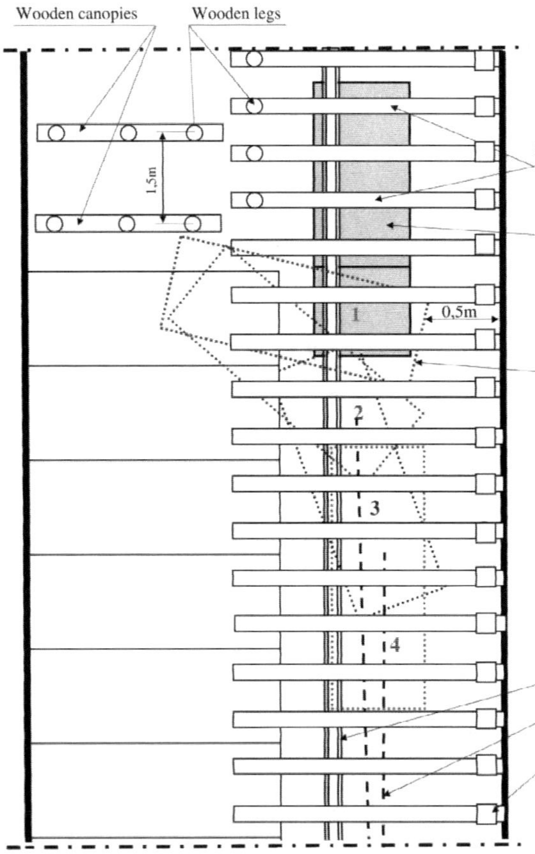

Figure 2. Powered roof support withdrawal with use of an insurance roof support.

Use of the insurance roof support for additional roof protection in the area of withdrawal of next roof supports as well as for separation from a roof fall zone of the liquidated cross-cut entry significantly improves the work safety. It also eliminates the necessity of setting and relocating the sprag for moving the roof support pulled out from a row.

<u>With use of devices intended for realization of powered roof support withdrawal and its positioning parallelly to the cross-cut entry</u>

UDW Ryś device developed and constructed in the Division for Powered Roof Supports in KOMAG [3], presented in Figure 3, is the device realizing the following functions: protection of the roadway's roof in the powered roof support withdrawal area, moving the roof support to the transportation section, positioning the roof support parallelly to the cross-cut entry as well as loading the roof support on the transportation platform.

The device consists of two parts connected with each other by the advancing rams which are moved against each other on guides. The first part consists of a base, on which 3 hydraulic legs are installed, 2 on left and 1 on right side. The second part consists of a base, on which 2 hydraulic legs are installed, one on each side. The legs are supporting the roof by through canopies. The legs are stabilized in the working position by the hydraulic cylinders. Connection of both parts by guides and hydraulic cylinders enables self-advancing of the device along the cross-cut entry.

IOP Conf. Series: Materials Science and Engineering **679** (2019) 012003 doi:10.1088/1757-899X/679/1/012003

Figure 3. UDW Ryś device for withdrawal of powered roof supports.

In the Polish mining industry, also UWS–PUMA device of similar design, manufactured by Sigma with its own advancing system or without it, was also used, and a hydraulic loading ramp type HRZ-2 [2].

The following methods for transportation of powered roof supports from cross-cut entry, used by mines can be distinguished:

- transportation of roof support on the floor (optionally on a special plate) using a winch placed in an upper entry;
  rope winches which are used have one disadvantage consisting in reduction of pulling force along with the progress of the rope winding. This phenomenon has adverse effect on the process of pulling the powered roof support. During rope transportation there is a high probability of rope break. Factors conducive to this may include mining and geological conditions (transverse and longitudinal wall inclination), crew selection, day of the week, technical condition of the equipment used for transport, etc. A broken rope may behave in an unpredictable manner, which has often been the cause of serious accidents.
- transportation of roof support by suspended monorail;
  the use of suspended monorail for the transportation of powered roof supports from the longwall to the main transportation roadway requires the construction of the railway route in the transport area, which is a time-consuming process and requires employment of additional crew during its construction. The use of suspended monorail for transporting roof supports from the longwall requires a greater height of the transportation field, because the height of the transported roof support additionally increases the height of the railway route and transportation beams. Transport of the powered roof support directly from the liquidated longwall to the new location by means of a suspended monorail, despite the initial large amount of work, significantly reduces the time of the entire transportation process.
- transportation of roof support by floor-mounted railway;
  use of floor-mounted railway for transportation of powered roof support directly from the cross-cut entry requires building the railway route on the cross-cut entry floor is a time-consuming process and requires employment of additional crew during its construction,

however it can reduce time required for dislocation of roof supports between the longwalls.

- transportation of the powered roof support on a sliding platform pulled by a rope of the winch along the troughs of partially disassembled face conveyor or along the trucks installed in the cross-cut entry;
  such transportation method has the same disadvantages as those in the case of roof support transportation on the floor (optionally on a special plate) using a winch rope.
- transportation of roof support using the self-propelled machines [4] [5];
  such a method of transportation does not require additional transportation equipment. However, due to size and turning radius of such machines, transportation space of bigger dimensions than in the case of above mentioned roof support transportation systems is required. Besides, in the case soft floor the machine can immerse into the floor. Each subsequent pass causes more and more wheel-ruts and driving problems.
- transportation of roof support using the chain capstan;
  a withdrawn roof support is pulled out to the transportation sledge connected with a pulling chain using the pulling harness and it is dragged from a longwall to upper entry. The chain drive is installed in a special stable in the upper entry and the chain return end is bolted ahead of insurance roof support using the expansion stands and lashings. Compared with the rope, the chain used to transport the roof support breaks much less frequently, and in the case of breaking it does not cause whipping as much as a broken rope. Changing the location of the return end, requiring it to be pulled up and bolted again with shortening the chain is a time-consuming operation in this transportation method.

Characteristics of the features of currently used technologies for withdrawal of the roof support from a row in a liquidated longwall as well as its transportation in the cross-cut entry, enabled developing the concept of the devices and procedures that make the both operation more effective. The devices designed in KOMAG based on this concept, using the virtual prototyping [6], are presented in chapter 3

3. A concept of the system for powered roof support withdrawal and transportation in the process of longwall panel liquidation

As a result of the analysis of the advantages and disadvantages of the technologies used so far, it was decided to use and combine the advantages of the UDW RYŚ device and the chain capstan. The device was designed based on the UDW RYŚ device, equipped with a fixation of the chain capstan return end and a special transportation plate. The system equipped with this device supports the process of withdrawal of a powered roof support from the moment of it's yielding in a row of roof supports standing in the cross-cut entry until the transportation of the roof support begins, using the basic mean of transportation in the upper entry. The system is shown in Figure 4 as an example of its use in cross-cut entry prepared for longwall liquidation.

The developed system consists of a device for roof support withdrawal and loading (A), of a chain capstan (B) as well as transportation plate (C).

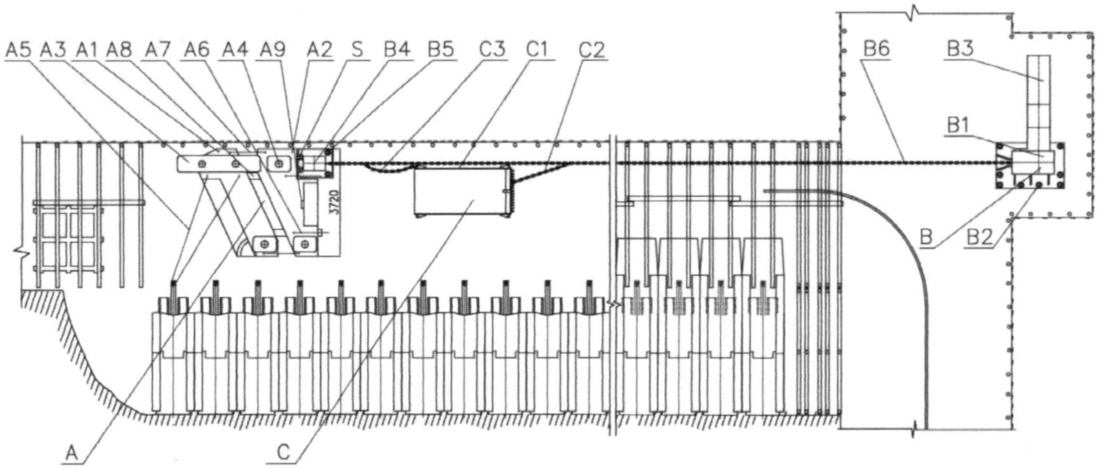

Figure 4. System for powered roof support withdrawal and transportation in the process of longwall panel liquidation.

Device for roof support withdrawal and loading on the transportation plate is a self-advancing unit consisting of the advancing member I (A1) and the advancing member II (A2). Both members are connected using the set of advancing rams (A7) as well as metal guiding sheets (A8) enabling mutual movement of the members keeping linearity of their movement. The set of advancing rams enables positioning the device in a such position, which makes installation of the next roof supports in a row in the cross-cut entry possible. The canopy (A3) is installed on the advancing member I supported by two hydraulic legs and the canopy (A4) is supported by one hydraulic leg. Hydraulic legs of the canopies are stabilized vertically using additional tilt (stabilising) cylinders. The advancing member I has a cant in a front part and a fender facilitating entering the roof support to the device transportation compartment. In the rear part of the advancing member I, permanently attached metal guiding sheets (A8) are used when the device is advancing. A set of two manoeuvring cylinders (A5) used to pull the powered roof support out of the row is placed on the side of the transportation section of the device. These cylinders are connected in an articulating and rotating way, which makes it possible to maneuver them in every direction. Two canopies (A4) are installed on the advancing member II, each is supported by a hydraulic leg. The canopy's hydraulic legs are also vertically stabilized by tilt (stabilising) cylinders. The advancing member II has a cant in the rear part to facilitate sliding the roof support off the device onto the transportation plate (C1). In the front part of the advancing member II there are guides through which the metal guiding sheets pass (A8). Next to the hydraulic legs, from the side of the transportation compartment, there is a set of two manoeuvring cylinders (A6) used to pull up, correct the position and support the transportation plate (C1). These cylinders are connected in an articulating and rotating way, which makes it possible to manoeuvre them in every direction. The advancing member II is equipped with a detachable chain catch (A9), which is used to pull up the roof support inside the transportation compartment of the device. The return end of a chain puller (B4) is connected to the rear part of the advancing member II via bolt connections (S). Canopies, cants and fenders of advancing members I and II define ends of the transportation compartment for the roof support. The withdrawn roof support is pulled into the transportation compartment, moved and pulled out onto the transportation plate. The canopies (A3 and A4) are equipped with articulating and rotating catches for installing additional manoeuvring cylinders, which will be fixed to transportation catches on the canopy or the gob shield of the withdrawn roof support via chains. They will additionally stabilize the position of the roof support while pulling (especially on inclination) and facilitate the rotation of the roof support during entering it into the transportation compartment of the device. The device for roof support withdrawal_and loading onto the transportation plate of the roof support can be

IOP Conf. Series: Materials Science and Engineering **679** (2019) 012003 doi:10.1088/1757-899X/679/1/012003

made in the left or right version depending on the location of the roof supports being withdrawn in the cross-cut entry.

Chain puller (B) consists of a drive end (B1), a return end (B4) and the main transportation chain (B6). The drive end (B1) consists of a body installed on a plate, in the centre of which a chain driving drum, positioned vertically and driven by a propulsion unit (B3) is placed. The drive unit includes a reduction gear, a clutch with a brake and an electric motor. The drive end is bolted using the hydraulic legs. The chain puller's return end (B4) is connected to the device withdrawing the roof support (A) by pins (S). The return end consists of a body installed on a plate, in the centre of which a return passive chain drum, also positioned vertically and additionally stabilized using 2 bolting legs, is placed. The main transport chain (B6) is a typical mining chain used in flight-bar conveyors. The bottom part of the chain moving at the floor is the driving part of the puller and the transportation plate (C1) is attached to it. The upper part of the chain is passive. The chain is made of easy-to-dismantle sections (quick connecting links) which is associated with the need to shorten it as more powered roof supports are withdrawn.

The complete transportation plate (C) consists of a main plate (C1) and detachable fastening chains, long one (C2) and short one (C3). The transportation plate (C1) is built in the form of a metal sheet vessel, open at the back. This is necessary to pull out the roof support. In the front part, the plate has a cant to facilitate its movement, and from below it is equipped with skids to facilitate dragging along the floor. Also in the front part, the plate is equipped with a series of catches, to which a long fastening chain (C2) is attached. These catches allow the transportation plate to be parallel to the main chain and to move it linearly when pulled with the roof support along the longwall cross-cut. The transportation plate (C1) in the front part, inside, is equipped with a handle for hooking the advancing system beam of the roof support when it is pulled onto the plate and the socket for installation of an additional bolting leg. The transportation plate is equipped with side covers protecting the transported roof support against contact with the passive line of the main transportation chain (B6).

4. The principles of operation of the devices included in the system for withdrawing and transportation of powered roof supports in the process of the longwall panel liquidation

In the initial state (Figure 5), the device for withdrawing and loading the roof support onto the transportation plate is positioned at the length of the last roof support to be removed from a row. The hydraulic legs are extended, the transportation plate is pulled to the device. Advancing segments of the device are pulled together to a minimum distance. The set of cylinders for roof supports withdrawal is ready for operation. The auxiliary cylinders are installed on a large canopy (supported by two legs). The roof support to be pulled out is yielded. The beam of the roof support advancing system is retracted (cylinder of the advancing system is extended to its maximum length). First, the advancing system of the pulled out roof support is connected with the device's set of cylinders. The cylinders are extended to the maximum length. Using these cylinders, the roof support is pulled out of a row in a few steps by shortening the chains. The withdrawn roof support must be moved to the wedge edge of the device's overrun segment to the length of the guiding fender (Figure 6).

Figure 5. Pulling out the roof support from a row in the cross-cut entry.

Figure 6. Pulling the roof support to the device transportation compartment.

Then, after disassembling the chains and roof support pulling out cylinders, the advancing system is connected with a special catch of the device sliding segment. The auxiliary cylinder on the canopy of the overrun segment supported by two legs is connected with the transportation catch on a gob shield by a chain (Figure 7). This cylinder enables rotation of the roof support in a direction of the device transportation tunnel. The roof support is rotated resting on the guiding fender, correcting at the same time its position with use the beam of its advancing system. Few times retraction of the cylinder with a chain shortening at the same time enables positioning the powered roof support in a front of the device transportation tunnel.

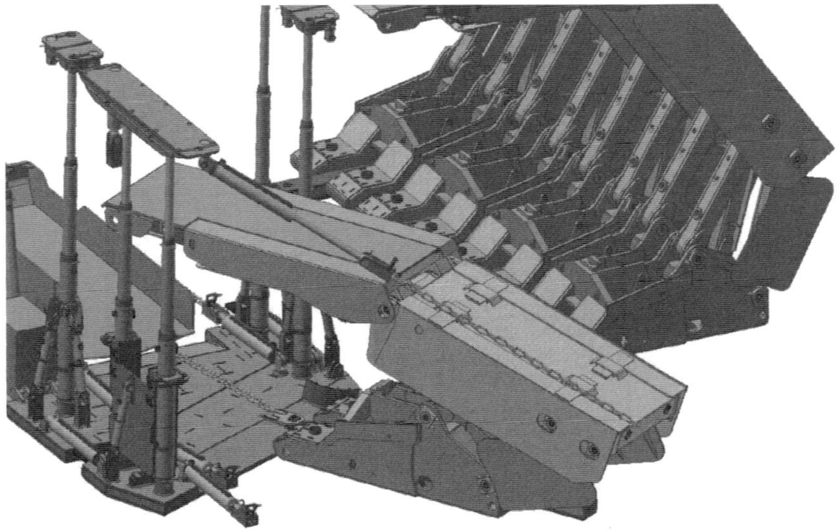

Figure 7. Pulling the powered roof support to the device transportation compartment.

The next step is to pull the roof support into the transportation compartment of the device. After unfastening the cylinder with the chain connecting the beam of the roof support advancing system with the eye of the sliding segment of the device and connecting the section beam with the chain with the eye in the transport plate, it is possible to pull the roof support into the transportation compartment of the device, using its advancing system (Figure 8). At the same time, the cylinders connecting the canopies of the device with the components of the roof support enable correcting its position and protecting it against tipping over.

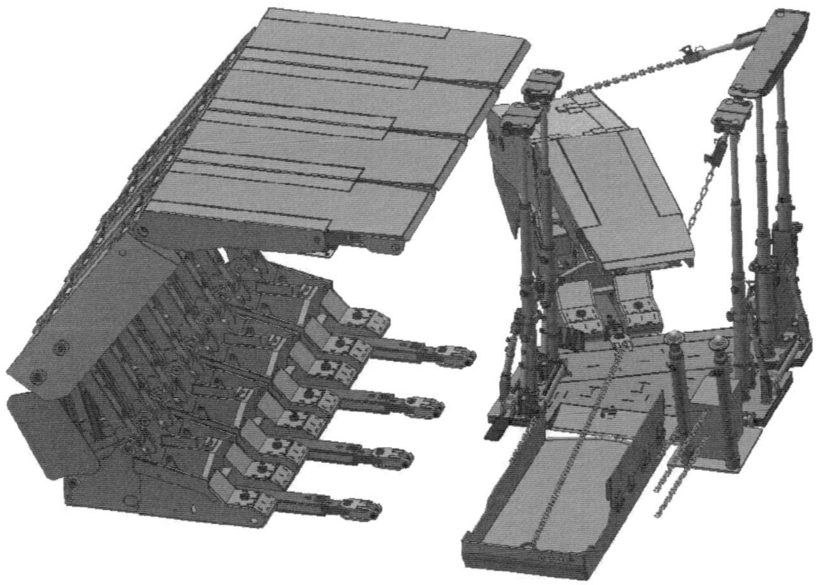

Figure 8. Pulling the powered roof support to the device transportation compartment using the transportation plate.

After shortening several times the chain connecting the beam of the advancing system of the roof support with the transportation plate holder, the roof support can be pulled onto the transportation plate using its advancing system (Figure 9).

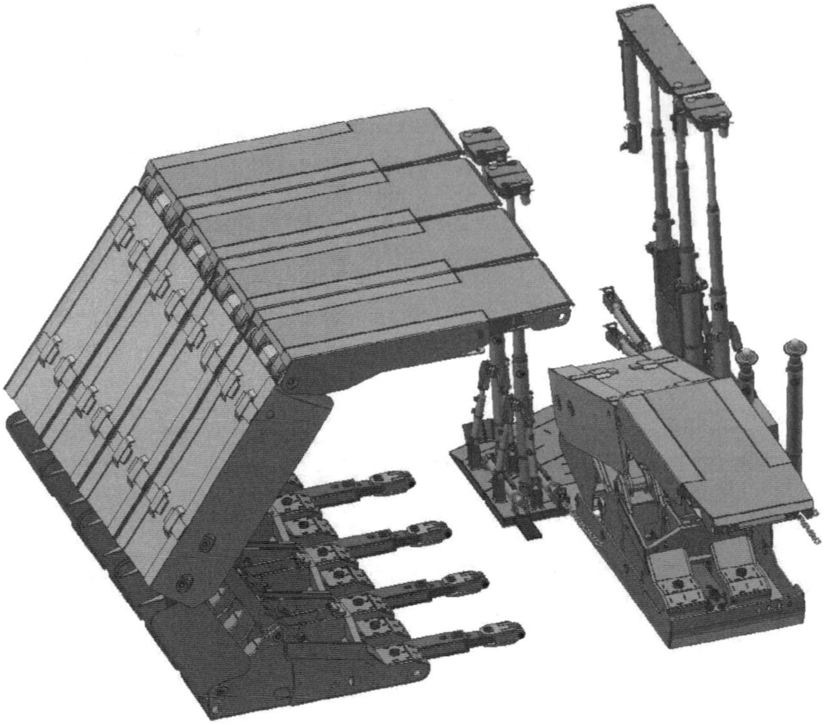

Figure 9. Powered roof support pulled onto the transportation plate.

In this position, the roof support is pulled along with the transportation plate to the tail gate, from where it can be transported by other means. The roof support with the transportation plate is pulled along the wall cross-cut entry by a chain puller, whose return end is connected to the basic device (Figure 10). The pulling process takes place according to the diagram shown in Figure 4.

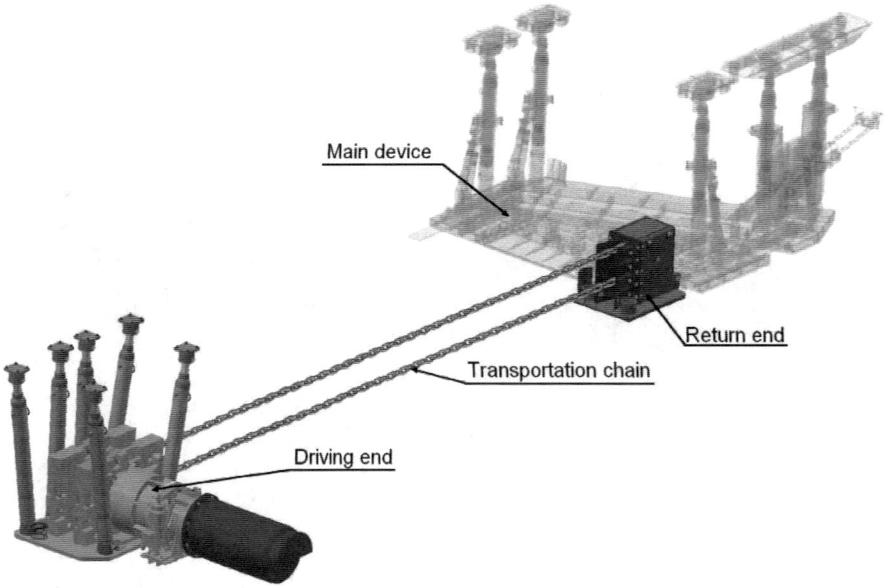

Figure 10. Chain puller.

The integration of the chain puller return end with the device for withdrawal and loading the roof support onto the transportation plate solved the basic problem when using this type of pulling machine in the process of liquidation of longwall equipment, i.e. repeated moving and anchoring of the return end. In the new solution, the anchoring force is transmitted to the return end by the device itself. The return end is also moved by the device advancing system. In the previous applications of the chain puller, the return end was pulled up using an additional sprag, cylinder and chain. Often, even 6 bolting legs did not provide a sufficient anchoring force and it was necessary to use additional chain lashings.

5. Conclusions

Based on the systems used for moving the powered roof support from the liquidated longwall panels, a new device was developed combining the advantages of the different existing solutions and enabling the easier movement of the powered roof supports in the liquidated longwalls in all its phases. The presented set of devices creates a system of work mechanization, which previously required much greater workload and more time. In the new, innovative solution, all cylinders used for moving the roof supports are ready for use at any time and there is no need to prepare the anchoring points for them at any stage of roof support movement. The work related to loading the roof support onto the transportation plate is also supported. The device also protects the roof of the longwall cross-cut entry in the area of subsequent withdrawal of roof supports, definitely improving the work safety. The device is self-advancing, what eliminates the need to use other means for its movement. A dedicated chain puller integrated with the device also improves safety due to the elimination of previously used rope winches.

References

[1] Nawrat S and Pytlik P: „Bezpieczna likwidacja ścian w kopalniach węgla w warunkach zagrożenia wybuchem metanu". *Bezpieczeństwo Pracy i Ochrona Środowiska w Górnictwie Nr 2 (2013)*, s. 3-12

[2] Stefaniak R, Janik M and Kuska J: „Nowe rozwiązania organizacyjno-techniczne wprowadzone przy przezbrajaniu kompleksu ścianowego ze ściany 904 do ściany 905 i zmiany wprowadzone do maszyn i urządzeń wyżej wymienionego kompleksu". *KOMTECH 2009, Innowacyjne, bezpieczne oraz efektywne techniki i technologie dla górnictwa Człowiek – Maszyna – Środowisko, Instytut Techniki Górniczej KOMAG, Gliwice 2009* s. 53-63; ISBN 978-83-60708-38-5

[2] Kania J, Szweda S and Szyguła M: „Analiza procesu wybudowy sekcji obudowy zmechanizowanej z wyrobiska ścianowego w aspekcie bezpieczeństwa". *Maszyny Górnicze Nr 3 (2018)*, s. 24-34

[4] Nikitina A M, Rib S V, Borzykh D M and Dadynsky R A: „Remounting of a mechanized complex using Petitto Mule equipment in the conditions of mines in the south of Kuzbass". *IOP Conf. Series: Earth and Environmental Science 206* (2018)

[5] Chase F, Worley P and McComas A.: "Longwall Shield Recovery Using Mobile Roof Supports". *Lakeview Scanticon Resort & Conference Center, Morgantown, WV, USA : July 31 - August 2, 2007*

[6] Winkler T and Tokarczyk J: Multi-Criteria Assessment of Virtual Prototypes of Mining Machines. *Proceedings of the World Congress on Engineering and Computer Science 2010 vol II. WCECS 2010, October 20-22, 2010, San Francisco, USA.* ISSN: 2078-0958 pp 1149-53.

IOP Conf. Series: Materials Science and Engineering **679** (2019) 012004 doi:10.1088/1757-899X/679/1/012004

Analysis of magnetic pull effect in disk induction motor with symmetrical/asymmetrical air-gap

T Wolnik[1]* **and B Będkowski**[1]

[1]Łukasiewicz Research Network – Institute of Electrical Drives and Machines KOMEL, Al. Roździeńskiego 188 40-203 Katowice, Poland

*E-mail: t.wolnik@komel.katowice.pl

Abstract. Disk induction motor presents an interesting alternative to standard machine on account of its limited axial dimension. Its application may be contemplated for mine industry machines. However, in case of the design presented in this paper, a most significant drawback lies in magnetic pull axial forces appearing during normal operation of the machine and affecting bearing performance. This is why such forces should be considered at the design stage; this is particularly true, when design of magnetic circuit does not counterbalance those forces. The issue of magnetic pull occurring in disk induction motor with a single air-gap is presented in the paper for both symmetrical and asymmetrical air-gap. Results of calculations and measurements of axial forces under differing operating conditions of the motor are demonstrated. Particular attention is paid to analysis of axial force waveforms and their variations for different motor operating conditions.

1. Introduction

Disk induction motors operate on the identical principle as standard cylindrical induction motors. The difference lies in the shape of magnetic circuit. In case of disk machines, axial dimension is limited at the expense of increased diameter. Such motors are therefore well adapted to applications, where one of main criteria is limitation of axial length of the machine [1, 2]. Examples of such applications may be found also in broadly defined mine industry.

Among designs of disk induction machines, we may distinguish a motor with a single air-gap, that is one stator and one rotor [1, 2, 3]. This design is justified in case of low power motors, where it is not necessary (and not acceptable from economical point of view) to use popular designs with double stator and double rotor [2]. A significant drawback of this design is that high magnetic pull axial forces (occurring between stator and rotor) emerge during normal operation of the machine [1, 2, 3, 4]. These forces substantially affect bearing performance and therefore should be taken into account during design of mechanical construction and selection of machine subassemblies [1, 2, 3]. Numerous publications on measurements and computational analyses of radial magnetic pull in cylindrical machines are known. Results of such research have been presented by Dorrell et al. [5,6], Tenhunen et al. [7] and others. However, there is a notable absence of publications dedicated to axial magnetic pull issue in disk motors, even though impact of this pull on machine operation is significant.

Results of calculations and experimental tests of magnetic pull axial forces are given in this paper. Investigation has been devoted to disk induction motor, its model in shown in Figure 1 and Figure 2.

The tests have been conducted at a dedicated test bench equipped with strain gauges EMS70-5kN. The main goal of the work was to determine values of magnetic pull axial forces present in the motor and their variations in selected motor operating conditions, to evaluate dominant frequencies in time courses of the axial forces and to determine the impact of magnetic pull on motor's structural elements.

2. Model of disk induction motor

During the research, we have investigated model disk induction motor AFIM11, with single stator and single rotor, number of pole pairs 2p=6. The basic data on geometric dimensions of magnetic circuit, winding and rated parameters of the discussed motor are shown in Table 1; the physical model of electromagnetic circuit elements is shown in Figures 1 and 2. The stator/rotor cores of model motor are made of a generator strip roll M470-50A, with slots cut out by electro-erosion (spark erosion) method. The stator slots are packed with symmetrical three-phase winding with winding factor equal to kw=0.933. Rotor's cage winding is fitted with two end-rings (upper and lower).

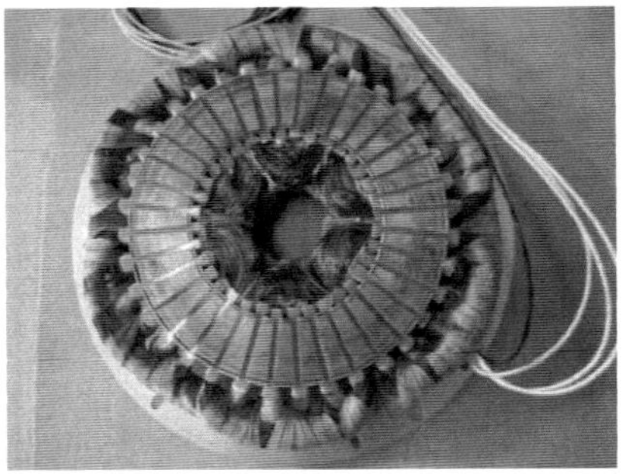

Figure 1. Stator with winding.

Figure 2. Rotor fitted with bearing into the shaft.

Table 1. Basic design data of model motor AFIM11.

Parameter			
Outer diameter - stator and rotor	Dz	[mm]	205
Inner diameter - stator and rotor	Dw	[mm]	130
Length of motor's magnetic circuit	L	[mm]	100
Air-gap length	δ	[mm]	0.85
Number of poles	2p	-	6
Number of stator phases	m	-	3
Number of stator slots	Qs	-	36
Number of rotor slots	Qw	-	40
Stator's winding factor	kw	-	0.933
Rated voltage	ULL	[V]	400
Rated frequency	fs	[Hz]	50
Rated power	Pn	[W]	1500
Rated torque	Tn	[Nm]	15
Rated current	In	[A]	3.4

Motor model shown in Figures 1 and 2 has been installed at a dedicated test bench, which had been adapted for measurements of axial forces of magnetic pull (Figure 3). The bench is equipped with four

IOP Conf. Series: Materials Science and Engineering **679** (2019) 012004 doi:10.1088/1757-899X/679/1/012004

strain gauges EMS70-5kN, located in the front panel; the stator together with its winding has also been secured to this panel. Placement and numbering of different sensors are shown in Figure 4.

Figure 3. Model of the test stand.

Figure 4. Placement of EMS70-5kN sensors - front panel of the test stand.

3. Magnetic pull forces during idle run

First of all, tests of motor with symmetrical air-gap equal to 0.85 mm (along the entire circumference) have been conducted. With motor running idle, the measurements of axial forces of magnetic pull versus supply voltage at constant frequency have been managed. The curve illustrating the dependence of resultant axial force of magnetic pull against the relative value of main magnetic flux is shown in Figure 5. Change of supply voltage at constant frequency produced changes in main magnetic flux Φ; in Figure 5 this flux has been related to the rated magnetic flux value Φ_n (at rated supply).

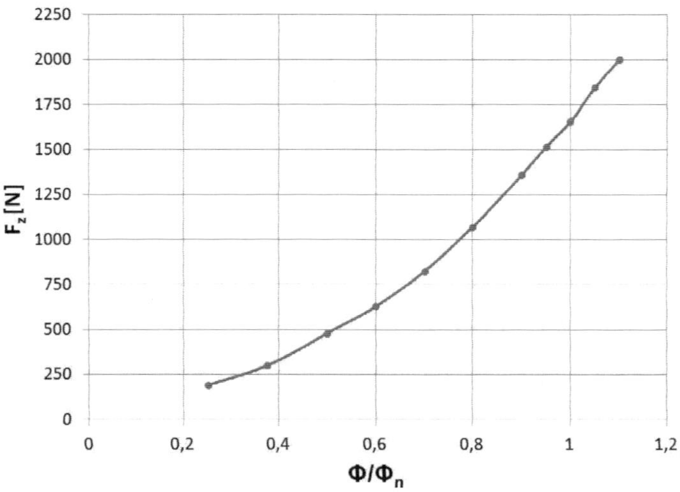

Figure 5. Dependence of axial magnetic pull force on main magnetic flux.

Attention must be drawn to measured value of magnetic pull axial forces for the rated value of main magnetic flux, this was measured at*c.* 1650 N. This is about 9 times greater than value of circumferential force F_r, which generates rotational torque; this force measured at averaged radii of stator/rotor disks is equal to *c.* 180 N. Methods for limiting such high axial forces are known [1, 2, 3];

however, in case of relatively low power motors their application would greatly increase the complexity of motor design and raise production costs. A cheaper solution of this problem is to strengthen the bearing assembly of disk machines so that such high axial forces could be withstood. This strengthening should be accomplished at the design stage.

For different values of supply voltage, corresponding to relative values of main magnetic flux as in Figure 5, the waveforms of axial force have been recorded and harmonic analysis has been conducted. The results are shown in Table 2 and Figure 6.

Harmonics with dominant amplitudes only are shown in Table 2. These are harmonics related to: magnetic field's rotational frequency (16.67 Hz) and its multiplicities and slot frequencies determined by rotor's rotational frequency and change in reluctance of magnetic circuit due to the presence of stator/rotor slots. The constant component has been removed for the sake of transparency from waveforms shown in Figure 6. The presented waveforms illustrate directly the pulsation value of magnetic pull axial force.

The recorded waveforms as well as the harmonic analysis show that predominant frequencies in the axial force waveforms are:

- magnetic field's rotational frequency(16.67 Hz),corresponding to synchronous speed of 1000 rpm,
- doubled frequency (100 Hz) of supply voltage; this has been explained thoroughly by Dorrel et al. [6],
- supply voltage frequency (50 Hz),
- doubled frequency (33.33 Hz) of magnetic field's rotation,
- characteristic frequencies 600-750 Hz due to changes in reluctance of magnetic circuit at a current rotor rotational speed.

As the main magnetic flux decreases, the amplitude of 16.67 Hz harmonic changes proportionately to change in main magnetic flux squared. In case of frequencies related to variable reluctance of stator/rotor magnetic core similar effect is observed; these are frequencies in the range 600-716.67 Hz. Their amplitudes change in proportion to change in main magnetic flux squared. We must also note the frequency 100 Hz (this is double of supply frequency fs=50 Hz). Amplitude of axial force at this frequency is nearly always dominant in all recorded waveforms, with variations in main magnetic flux. This is due to the mutual action of magnetic flux density and phase current [6].

Table 2. Pulsation amplitudes of axial force F_z for selected harmonic frequencies, when main magnetic flux values are varied and supply voltage frequency is constant and equal to f=50Hz; motor running idle.

Frequency of the pulsation F_z [Hz]	16.67	33.33	50	100	300	616.67	666.67	716.67
Amplitude of the pulsation F_z [N] Φ/Φ_n =1	7.76	1.27	1.65	7.53	2.26	2.8	0.39	1.11
Amplitude of the pulsation F_z[N] Φ/Φ_n =0.8	5.62	1.08	1.26	9.47	0.3	1.75	0.17	0.59
Amplitude of the pulsation F_z[N] Φ/Φ_n =0.6	2.64	0.91	0.64	7.34	0.29	1.04	0.13	0.41
Amplitude of the pulsation F_z[N] Φ/Φ_n =0.4	0.76	1.3	0.23	5.24	0.12	0.53	0.07	0.21
Amplitude of the pulsation F_z[N] Φ/Φ_n =0.2	0.31	1.18	0.5	2.29	0.1	0.21	0.03	0.1

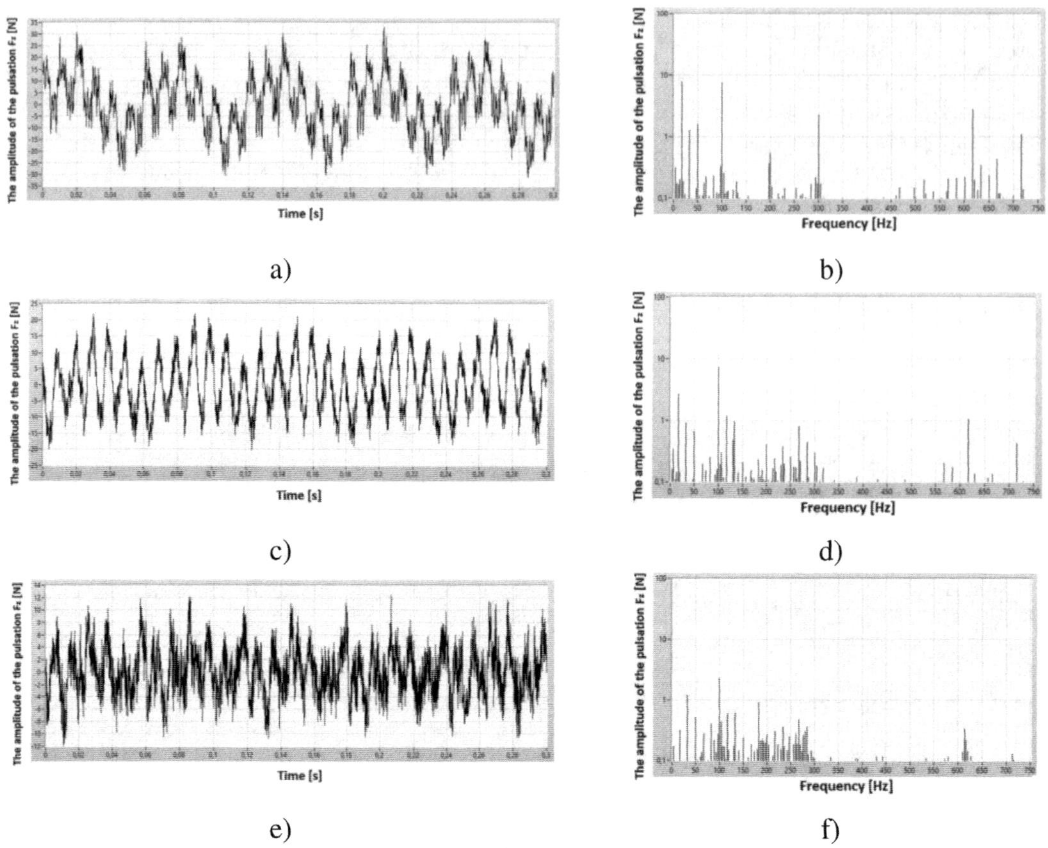

Figure 6. Waveforms and harmonic analysis of variable component of magnetic pull axial force; main magnetic flux is varied: a, b) supply400V,50Hz, Φ/Φn=1; c, d) supply 240V,50Hz; Φ/Φn=0.6; e, f) supply 100V,50Hz, Φ/Φn=0.2.

4. Magnetic pull forces at varying load

The waveforms and values of magnetic pull axial force have also been recorded with loaded motor and symmetrical air-gap. Motor was loaded with torque varying from zero (idle run) up to 22.5 Nm (this, corresponds to 1.5 times rated load torque). The waveforms with constant component removed are shown in Figure 7. Amplitudes of different dominant harmonic frequencies are set out in Table 3.

When these data are analysed, we may conclude that change in load torque of induction disk motor leads to following effects in axial force pulsation waveforms:

- decrease in amplitude value of 16.67 Hz harmonic (which is due to magnetic field's rotational speed),
- increase in amplitude value of 100 Hz harmonic (this is 2fs frequency),
- shift of harmonic components (due to changes in reluctance of magnetic circuit and current rotor's rotational speed) towards lower frequencies (this is similar to decrease of rotor's speed corresponding to increased load). In case of motor's idle run the dominant reluctance frequency was 616.67 Hz, amplitude 2.8 N, while for load torque 20 Nm the dominant frequency was 573.33 Hz amplitude 2.9 N (Figure 7 b, f).

Moreover, increase of motor load leads to decrease of total average value of magnetic pull forces. This is shown in Figure 8.

Table 3. Pulsation amplitudes of axial force F_z for selected harmonic frequencies and average value of axial force ΣF_z at rated supply 400V, 50 Hz; varied motor load torque.

Frequency of the pulsation F_z [Hz]	**16.67**	**33.33**	**50**	**100**	**300**	**616.67**	**666.67**	**716.67**	**ΣF_z [N]**
Idle run	7.76	1.27	1.65	7.53	2.26	2.80	0.39	1.11	1656
2.5 Nm	5.89	2.02	1.30	8.26	0.24	0.62	0.25	0.19	1599
5 Nm	3.61	1.77	1.28	8.40	0.24	0.28	0.31	0.11	1573
7.5 Nm	3.63	2.33	2.58	9.47	0.46	0.52	0.10	0.05	1538
10 Nm	2.83	2.69	1.07	10.76	0.09	0.90	0.12	0.31	1504
12.5 Nm	2.01	2.47	0.43	11.59	0.30	0.68	0.06	0.29	1464
15 Nm	1.91	2.44	1.12	13.68	0.01	0.21	0.06	0.27	1408
17.5 Nm	1.78	1.81	0.38	15.16	0.08	0.04	0.19	0.34	1349
20 Nm	1.38	1.18	1.41	17.35	0.41	0.06	0.34	0.99	1286
22.5 Nm	1.34	0.60	3.2	20.18	0.13	0.26	0.28	0.14	1195

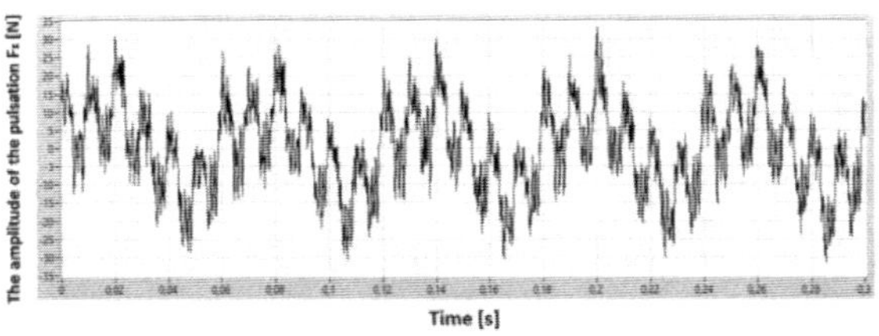

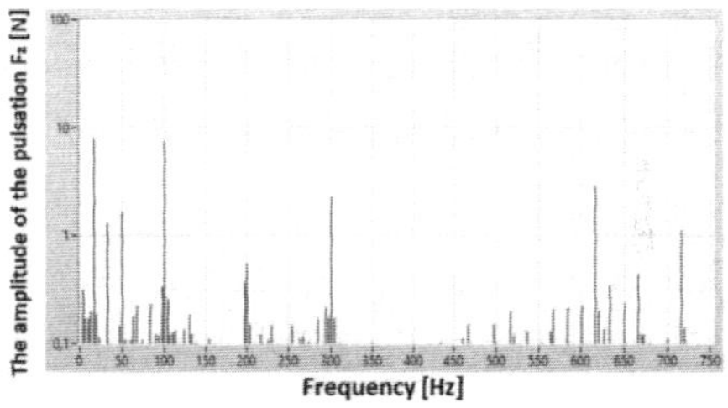

a) b)

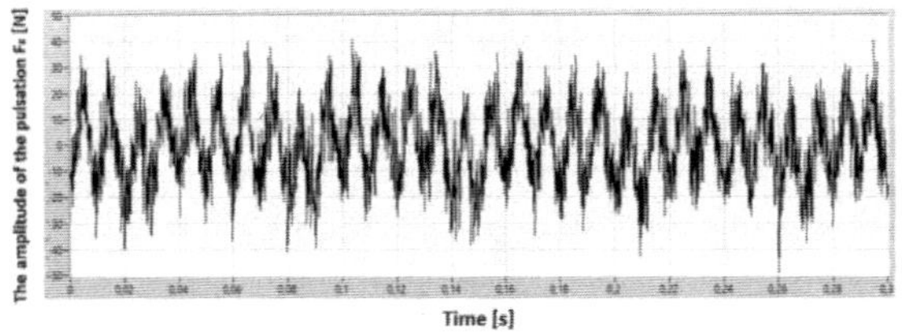

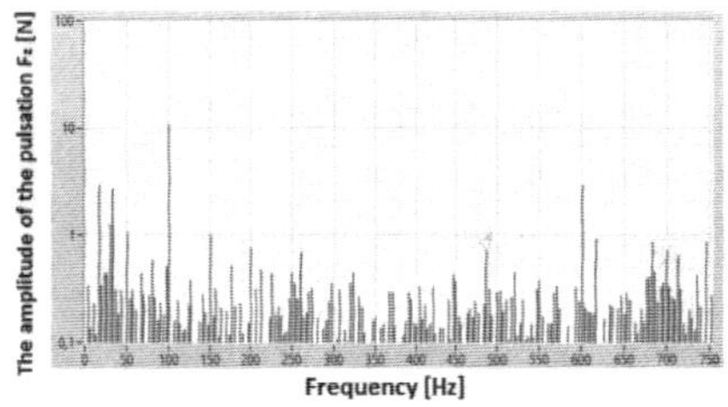

c) d)

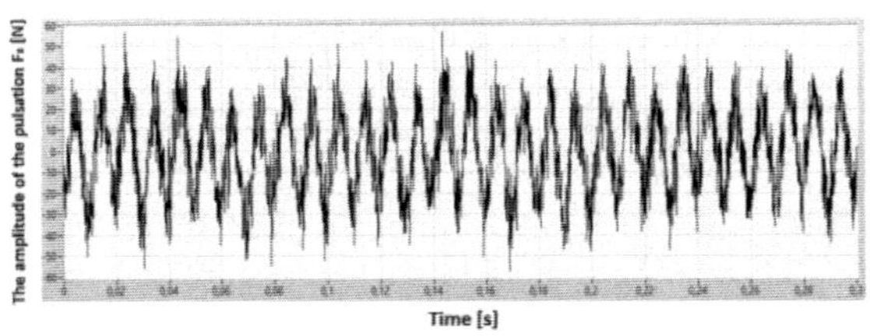

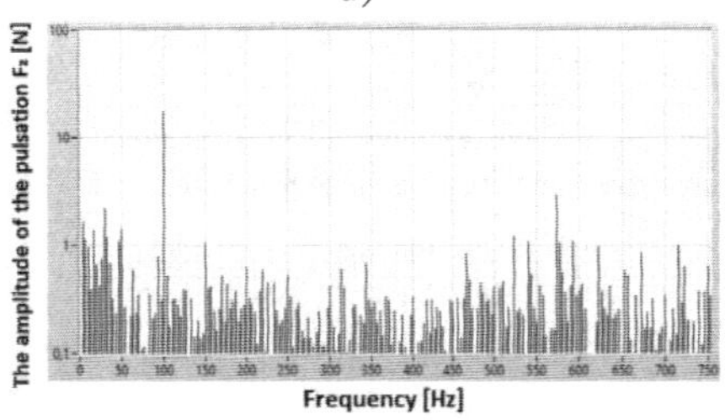

e) f)

Figure 7. Waveforms and harmonic analysis of variable component of magnetic pull axial force; rated supply 400V, 50 Hz, load is varied: a, b) idle run , c, d) load torque 10 Nm, e, f) load torque 20 Nm.

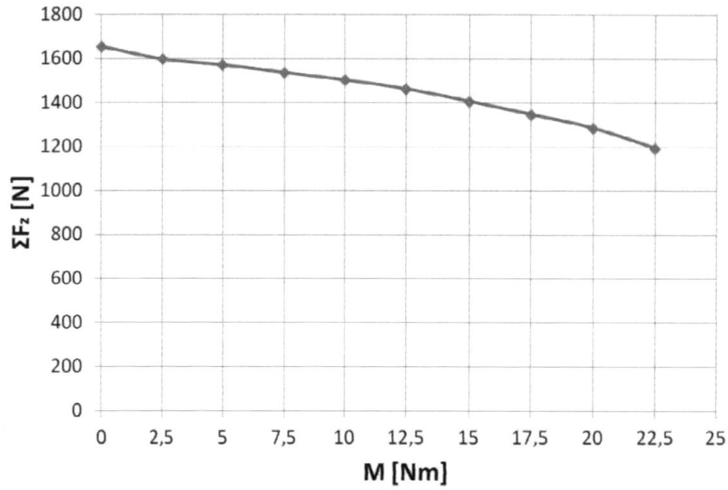

Figure 8. Dependence of magnetic pull axial force on motor's load torque.

5. Impact of air-gap asymmetry on magnetic pull forces

The problem of maintaining symmetry of the air-gap and circumferential symmetry in particular often arises in disk motors (Figure 9). This is caused by relatively large external diameters of rotor disks in relation to their thickness [1, 2]. If rotor design is faulty, then rotor may be whippy and its rigidity may be low. This asymmetry may also result from tolerance chain of different subassemblies. The length of the air-gap δ with circumferential asymmetry as shown in Figure 9 is described by relationship (1). This is a function of current disk radii r (where we analyse the air-gap), angle β and angle α measured along the circumference of r radius.

$$\delta(r,\alpha,\beta) = \delta_0 + r \cdot sin(\beta)\cos(\alpha) \quad (1)$$

where:
r – current disk radius,
α – current angle measured along the circumference of r radius, and:
β – angle of air-gap non-uniformity (disk angle of inclination)
δ0 –rated air-gap (under symmetrical conditions); with asymmetry this is an air-gap in shaft axis.

The dedicated test bench displayed in Figure 3 makes it possible to introduce circumferential asymmetries of the disk motor's air-gap. Therefore, impact of circumferential asymmetry of the air-gap on values and waveforms of magnetic pull axial forces in a model motor has been investigated. The tests have been run for two cases of circumferential asymmetry of the air-gap:

1. asymmetry Δ=0.5 mm, where δ0 = 0.85 mm, δmin = 0.5 mm, δmax = 1.0 mm
2. asymmetry Δ=1.0 mm, where δ0 = 0.85 mm, δmin = 0.3 mm, δmax = 1.3 mm.

The elaborated results are shown in Tables 4, 5 and Figure 10.

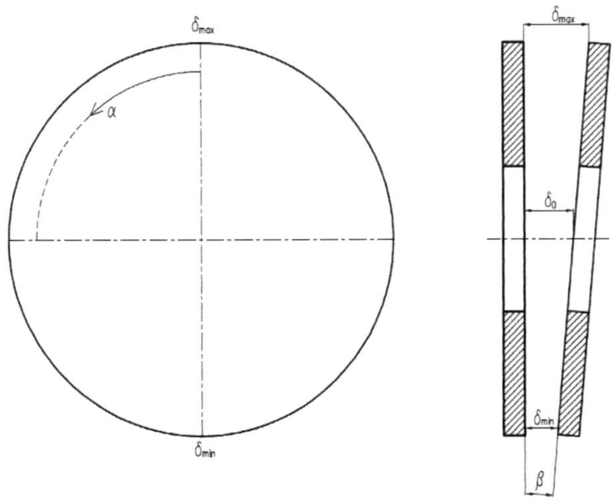

Figure 9. Circumferential asymmetry of the air-gap.

Results of measurements of average values of axial forces with different strain gauges (cf. Figure 4) are presented in Table 4. These results demonstrate the direction (geometrical) of evolving asymmetry; location of different strain gauges shown in Figure 4 is taken into account. When asymmetry is changed, values of forces measured by sensors F1 and F4 decrease, while those measured by sensors F2 and F3 increase; this testifies unequivocally to the direction of disk inclination. Appropriate placement of strain gauges may therefore constitute the basis for diagnosing air-gap uniformity in disk motors. As the air-gap asymmetry increases (disk inclination angle is greater), the total value of axial force ΣF_z defined by relationship (2) increases. However, these changes are not significant and are equal to *c.* 5%.

$$\sum F_z = F1 + F2 + F3 + F4 \tag{2}$$

Table 4. Average values of axial forces versus changes in circumferential symmetry of the air-gap; forces recorded with strain gauges F1-F4 at rated supply 400V, 50Hz, motor running idle.

	F1 [N]	**F2 [N]**	**F3 [N]**	**F4 [N]**	**ΣF_z [N]**
Circumferential symmetry of air-gap	417	416	379	444	1656
Circumferential asymmetry Δ=0.5 mm	347	468	461	394	1671
Circumferential asymmetry Δ=1.0 mm	308	519	529	383	1741

Comparison of waveforms and different harmonics of axial forces under conditions of symmetry and asymmetry of the air-gap is shown in Table 5 and Figure 10.

Table 5. Pulsation amplitudes of axial force F_z for selected harmonic frequencies versus changes in circumferential asymmetry of the air-gap;, main magnetic flux is varied, supply frequency is constant f=50Hz, motor running idle.

	Frequency of the pulsation F_z [Hz]	**16.67**	**33.33**	**50**	**100**	**300**	**616.67**	**666.67**	**716.67**
400V, 50 Hz	Symmetrical air-gap	7.76	1.27	1.65	7.53	2.26	2.80	0.39	1.11
	Asymmetrical air-gap Δ=0.5 mm	26.07	1.25	1.61	9.64	0.52	2.47	0.41	0.76
	Asymmetrical air-gap Δ=1.0 mm	56.05	2.04	1.55	9.49	0.15	2.81	0.71	1.3

IOP Conf. Series: Materials Science and Engineering **679** (2019) 012004 doi:10.1088/1757-899X/679/1/012004

320V, 50Hz	Symmetrical air-gap	5.62	1.08	1.26	9.47	0.3	1.75	0.17	0.59
	Asymmetrical air-gap Δ=0.5 mm	15.04	0.86	1.25	8.12	0.06	1.56	0.18	0.42
	Asymmetrical air-gap Δ=1.0 mm	31.39	0.78	0.38	9.95	0.24	1.96	0.36	0.96
240V, 50Hz	Symmetrical air-gap	2.64	0.91	0.64	7.34	0.29	1.04	0.13	0.41
	Asymmetrical air-gap Δ=0.5 mm	7.20	0.64	0.39	4.99	0.29	1.04	0.11	0.28
	Asymmetrical air-gap Δ=1.0 mm	12.91	0.18	0.64	5.64	0.22	1.07	0.19	0.61
160V, 50Hz	Symmetrical air-gap	0.76	1.3	0.23	5.24	0.12	0.53	0.07	0.21
	Asymmetrical air-gap Δ=0.5 mm	2.46	0.85	0.14	2.20	0.07	0.52	0.06	0.11
	Asymmetrical air-gap Δ=1.0 mm	3.87	0.44	0.49	2.62	0.06	0.61	0.07	0.32

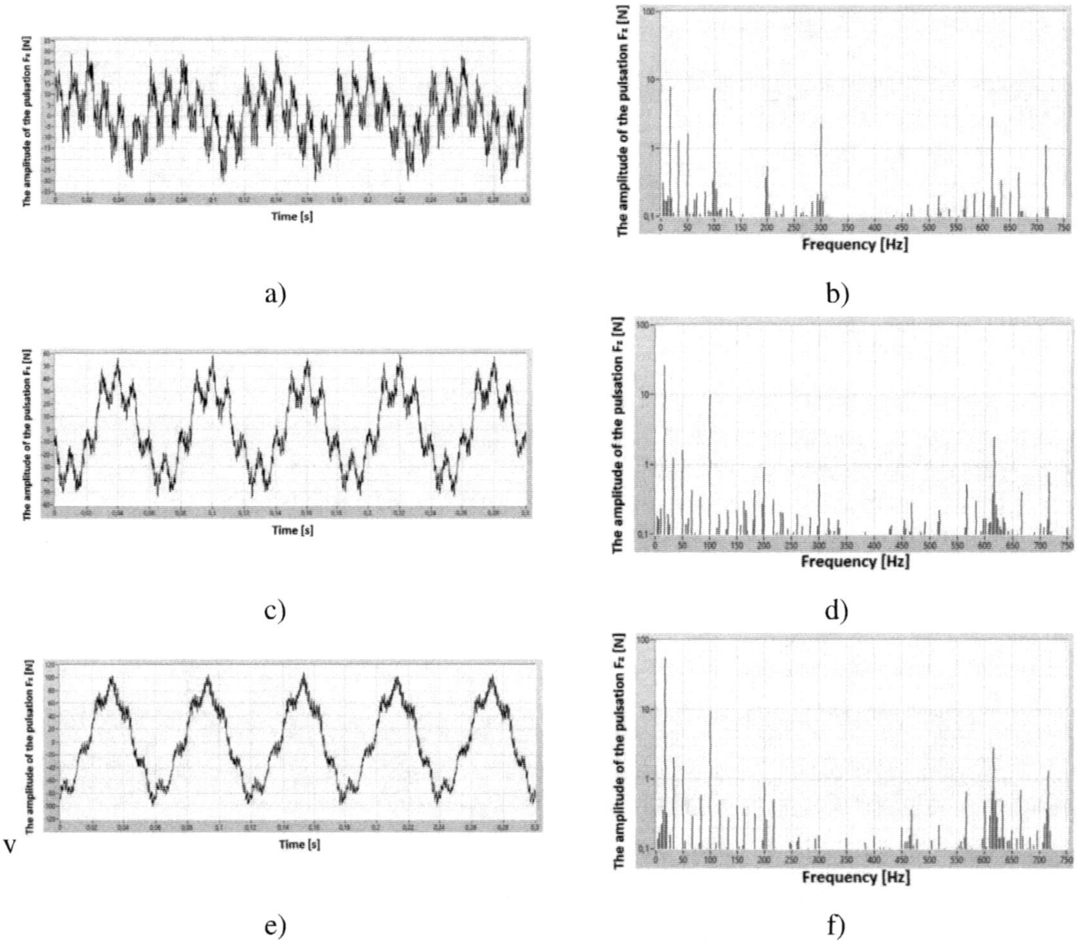

Figure 10. Waveforms of variable component and harmonic analysis of magnetic pull axial forces; circumferential asymmetry of the air-gap, rated supply 400V, 50Hz: a, b) for symmetrical air-gap; c, d) for asymmetrical air-gap Δ=0.5 mm; e, f) for asymmetrical air-gap Δ=1.0 mm.

On the basis of the results shown in Table 5 and Figure 10 it may be concluded that increase in the circumferential asymmetry of the air-gap leads to significant increase of the fundamental harmonic

IOP Conf. Series: Materials Science and Engineering **679** (2019) 012004 doi:10.1088/1757-899X/679/1/012004

component due to magnetic field's rotational speed (16.67 Hz), this is irrespective of the magnetic flux. This is also the reason for increased pulsations in the entire waveform of magnetic pull axial force. In case of remaining harmonics, it may be stated that circumferential asymmetry of the air-gap bears no impact on their amplitudes, and changes in these amplitudes are random. Therefore, if we want to diagnose the symmetry of the air-gap on the basis of axial force waveform, observation and analysis of the frequency component due to magnetic field's rotational speed is the only information required.

6. Impact of asymmetry and magnetic pull forces on structural elements – numerical computations

In rotational machines, the main sources of rotor load are torsional moment, centrifugal force and force of magnetic pull, which occurs between rotor and stator surfaces.It has been demonstrated earlier, that distribution of magnetic pull force depends largely on uniformity and length of the air-gap between rotor/stator surfaces; in case of disk machines this force may attain significant values. Therefore, it is reasonable to take this load into account in mechanical calculations and to assess its impact on motor's structural elements. It may be observed that most rotational machines operate under steady conditions. The variable conditions transpire usually when machine load or rotational speed changes; this happens most often during start-up or coasting. This is true in case of disk machines as well. When machine operates under variable conditions, the variable input functions appear; source of these inputs may be traced back to the machine itself, and the input forces are generated by rotating masses of residual unbalance of the rotor or they may be caused by varying distribution of magnetic pull forces due to e.g. circumferential asymmetry of the air-gap [8,9]. Hence, principal components of the input frequencies correspond to rotor's rotational frequency and its harmonics. The appearance of destructive vibrations is related to resonance, which will manifest itself at the instant, when rotational frequency of the rotor is identical as natural frequency of the rotor-bearings-support system [8,9]. This frequency is also termed rotor's critical frequency (resonance frequency). This frequency must be avoided; thus, natural frequencies of the rotor must be checked and machine design or rotational speed must be corrected, if necessary.

Examples of numerical analyses' results (static and modal) for model machine AFIM11 are shown below; identical machine was previously tested. These analyses have been run on the basis of simplified 3D rotor models. We have used module for simple mechanical analysis, which is available in Autodesk Inventor Professional software. The restraints corresponding to rotor bearing and shaft neck torque output are shown in Figure 11. The applied load produced by rotational torque generated in the machine and magnetic pull is shown in Figure 12; magnetic pull has been previously recorded during tests of model machine with the help of four strain gauges EMS70-5kN. Mechanical strength of the rotor in the single stator machine (AFIM11) has been analysed assuming different settings of air-gap uniformity (different values of magnetic pull force at the circumference, corresponding to measured values).

Figure 11. Boundary conditions (restraint)

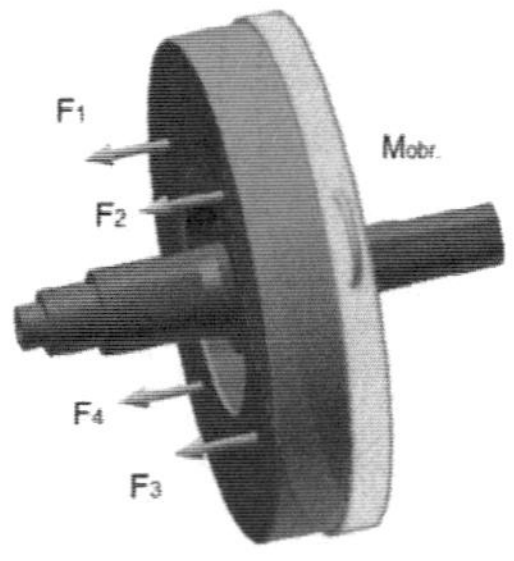

Figure 12. Boundary conditions (load torque Mo brand magnetic pull force F1-4)

Results of analysis for different load configurations are shown in Figures 13-16. These results contain information on reduced stress, dislocation and safety factor present and specific to constructional materials applied in rotor shaft and structural elements of rotor core.

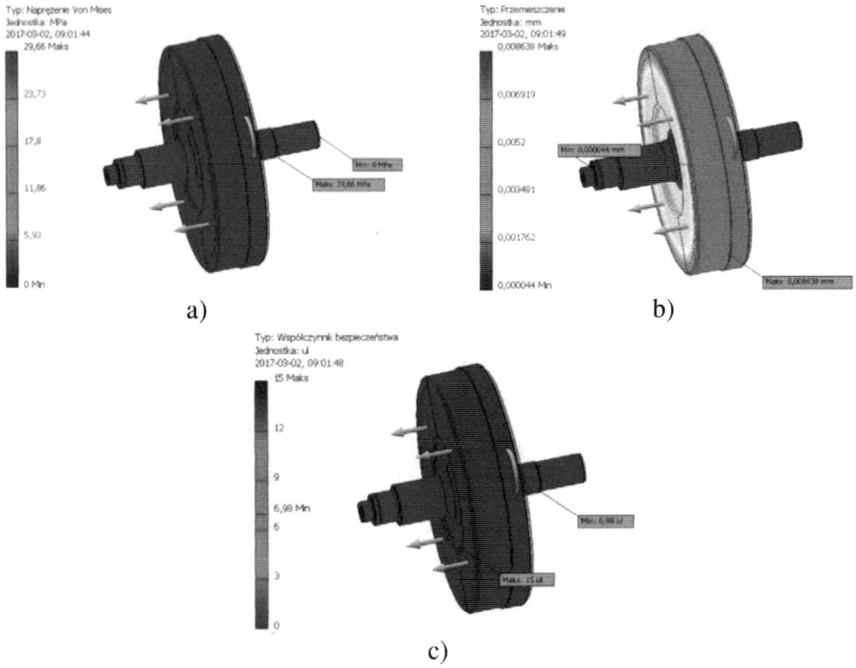

Figure 13. Results of static analysis, motor loaded with torque15 Nm and magnetic pull force F1-F4=350N; a) reduced stress, b) dislocations, c) safety factor.

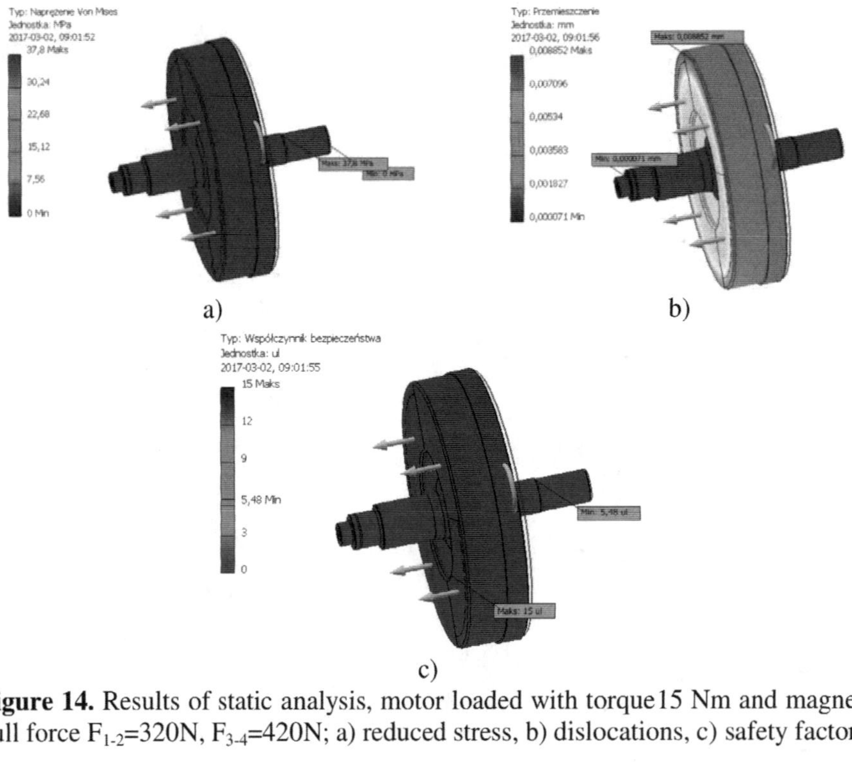

Figure 14. Results of static analysis, motor loaded with torque15 Nm and magnetic pull force $F_{1\text{-}2}$=320N, $F_{3\text{-}4}$=420N; a) reduced stress, b) dislocations, c) safety factor.

IOP Conf. Series: Materials Science and Engineering **679** (2019) 012004 doi:10.1088/1757-899X/679/1/012004

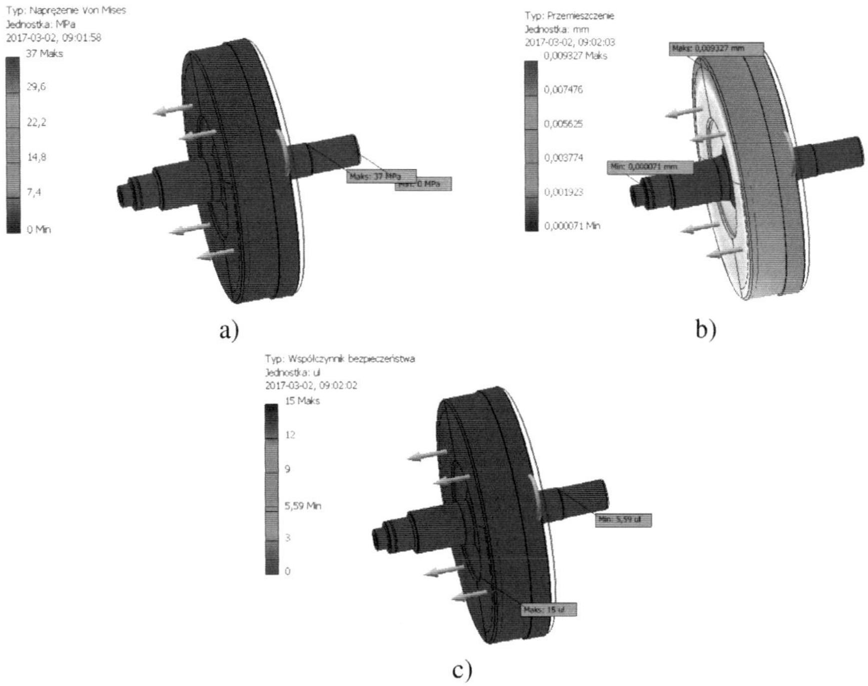

Figure 15. Results of static analysis, motor loaded with torque15 Nm and magnetic pull force $F_{1\text{-}2}$=250N, $F_{3\text{-}4}$=490N; a) reduced stress, b) dislocations, c) safety factor.

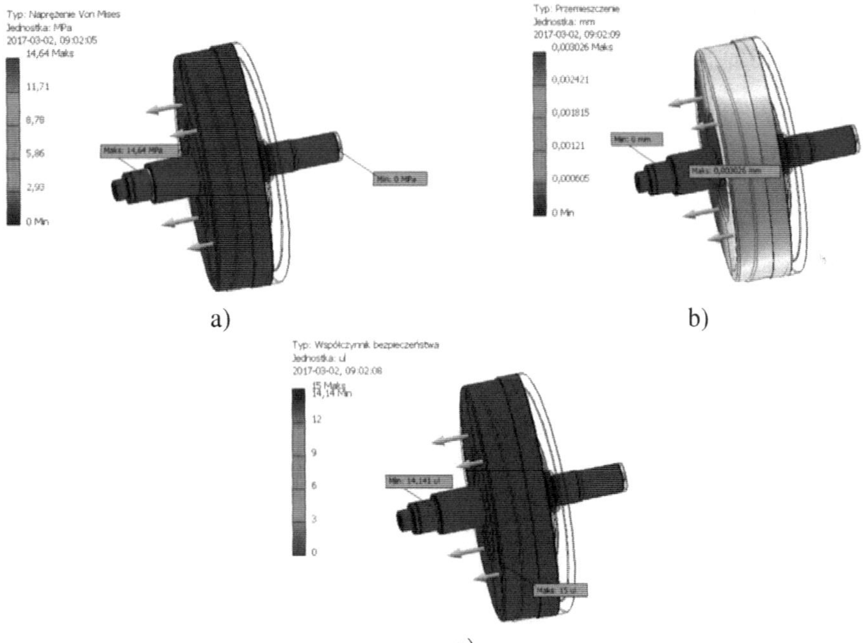

Figure 16. Results of static analysis, motor loaded with torque0 Nm and magnetic pull force $F_{1\text{-}2}$=250N, $F_{3\text{-}4}$=490N; a) reduced stress, b) dislocations, c) safety factor.

When results of simulations are analysed, it is clearly seen that stresses independently of assumed loads do not pose any hazard to rotor construction, and minimum safety factor for S235JR exceeds 5. Rotor construction is fairly rigid and rotor dislocation in case of analysed load does not exceed

IOP Conf. Series: Materials Science and Engineering **679** (2019) 012004 doi:10.1088/1757-899X/679/1/012004

0.01 mm. A problem may arise when load changes dynamically during motor rotation; at higher values it may influence the fatigue life of the rotor. The changing load generated by magnetic pull introduces a bending moment, which pulsates in accordance with rotational speed; in extreme cases this may lead to wear damage of rotor elements.

Results of modal analysis for six initial forms of natural frequencies are shown in Figure 17. The first critical speed of the rotor is equal to *c.*49 000 rpm; this is well outside machines peed range and there is no risk of resonance due to this critical speed.

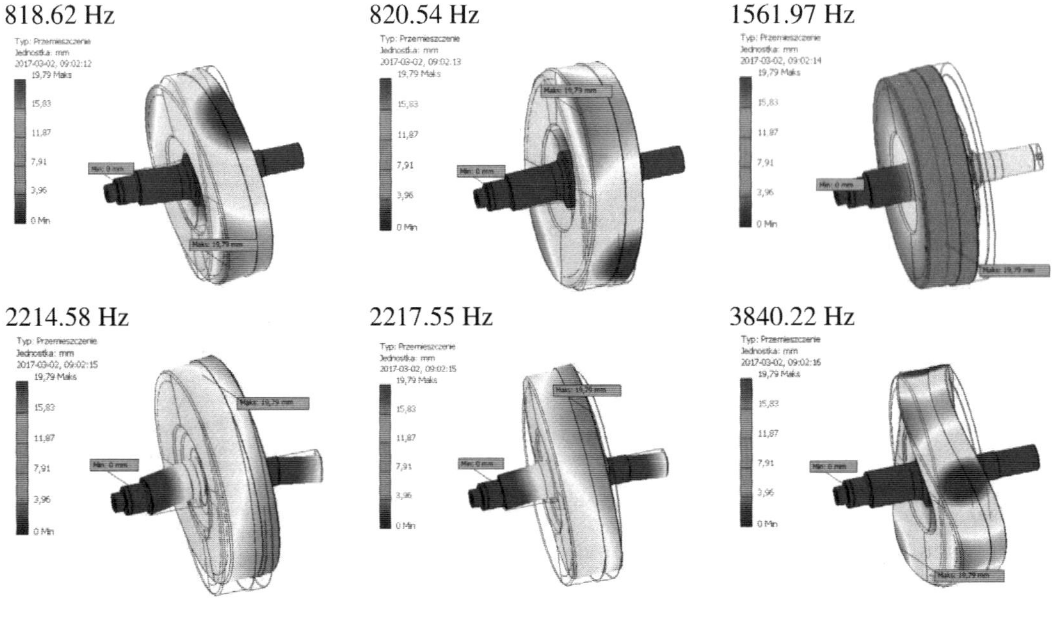

Figure 17. Results of modal analysis – vibration forms (6 initial natural frequencies).

7. Conclusions

Results of calculations and tests of magnetic pull effect and asymmetry of the air-gap in disk induction motor with a single air-gap are presented in this paper. The results make it possible to determine values of magnetic pull axial force under different operating conditions as well as its changes accompanying variations in supply or load parameters. For all analysed motor operating conditions, the pulsation waveforms and their harmonic analysis are given. Hence, it was possible to determine the characteristic frequency components and their changes relative to changes in input quantities.

The presented data should be useful to machine designers working on bearing assemblies. The incorrect calculation of axial forces may lead to decreased machine efficiency and premature wear of bearings due to faulty selection. This may be traced back to unfavourable friction effects and excessive mechanical losses. These forces should be taken into account, when equivalent load of the bearings is calculated. The dependence between bearing durability and bearing load is exponential in character. If we neglect the axial forces. then bearings' lifetime would decrease at least 3.33s times in relation to prognosed lifetime (s denotes the ratio of equivalent load to radial load). Hence, in such machines use of ball or angular contact cone bearings is recommended.

Acknowledgments

The project has been financed by the Narodowe Centrum Nauki (National Science Centre) within the framework of the research grant No. UMO 2012/07/B/ST8/04099.

References

[1] Parviainen A *Design of axial-flux permanent magnet low-speed machines and performance comparison between radial-flux and axial-flux machines.* PhD thesis, Lappeenranta University of Technology, 2005

[2] Valtonen M *'Performance characteristics of an axial-flux solid-rotor-core induction moto'*. PhD thesis, Lappeenranta University of Technology, 2007

[3] Nasiri-Gheidari Z and Lesani H A Survey on Axial Flux Induction Motors, *Electrical Review*, 2012, **2**, pp 300-305

[4] Wang R-J, Kamper M.J, Van der Westhuizen K, Gieras F.J: Optimal Design of a Coreless Stator Axial Flux Permanent-Magnet Generator, *IEEE Transactions on Magnetics*, 2005, **1**, (41), pp 55-64

[5] Dorrell D.G: Sources and Characteristics of Unbalanced Magnetic Pull in Three-Phase Cage Induction Motors With Axial-Varying Rotor Eccentricity, *IEEE Transactions on Industry Applications*, 2011, **1**, (47), pp 12-24

[6] Dorrel, D.G, Popescu M, Ionel D.M: Unbalanced Magnetic Pull Due to Asymmetry and Low-Level Static Rotor Eccentricity in Fractional-Slot Brushless Permanent-Magnet Motors With Surface-Magnet and Consequent-Pole Rotors, *IEEE Transactions on Magnetics*, 2011, **7**, (46), pp 2675-2685

[7] Tenhunen A, Benedetti T, Holopainen T P and Arkkio A Electromagnetic forces of the cage rotor in conical whirling motion, *IEE Procedings - Electric Power Applications*, 2003, **5**, (150), pp 563-568

[8] MadejJ, Sitek M: Modal and frequency analysis of electrovibrator kit for different configuration of activators, *Acta Mechanica et Automatica*, Vol. **5**, No.3, 2011, pp 87-90

[9] Będkowski B The structure analysis of the permament magnets electro vibrator model, *Maszyny Elektryczne: Zeszyty Problemowe*, Vol. **122**, 2019, pp.71-78.

IOP Conf. Series: Materials Science and Engineering **679** (2019) 012005 doi:10.1088/1757-899X/679/1/012005

Measurement of the pick holders position on the side surface of the cutting head of a mining machine with the use of stereoscopic vision

A Jagieła–Zając[1]* and P Cheluszka[1]

[1]Silesian University of Technology, Faculty of Mining, Safety Engineering and Industrial Automation, 2 Akademicka Street, 44–100 Gliwice, Poland

*E–mail: Amadeus.Jagiela–Zajac@polsl.pl, Piotr.Cheluszka@polsl.pl

Abstract. The efficiency of rock cutting with mining machines is largely determined by the arrangement of picks, i.e. the number and their arrangement on the working unit of the mining machine. Not only the correct selection of the pick system for given conditions at the design stage is important, but also ensuring compliance with the design of the finished product. Strives, among others therefore, for robotisation of the process of manufacturing cutting heads/drums. From the point of view of the robotisation of the pick holders welding process, it is necessary to assess in real time the position of the pick holders relative to the side surface of the cutting head body. A convenient way is to use contactless measurement methods based on vision systems. The article presents a method of determining the position of pick holders relative to the side surface of the cutting head body of a roadheader, during their positioning, using a 3D vision system. Data processing was carried out in the Matlab software using the libraries of the Computer Vision Toolbox. A mathematical model describing the transformation of images recorded by cameras has been presented. On the basis of this model, the distribution of distances between the pick holder base points and the side surface of the cutting head was determined for a given pick holder setting. The developed measurement method was tested on an experimental stand built in the Laboratory of robotics of the Department of Mining Mechanization and Robotisation at the Silesian University of Technology.

1. Introduction

To ensure high accuracy and repeatability in the manufacturing of cutting heads/drums of mining machines, robotisation of this process is sought [1, 2]. Cutting heads/drums consist of a body and pick holders welded to the side surface of this body, in which the picks are placed. The spatial position of the pick–pick holder system is determined by the values of six parameters determined at the design stage. The bodies of the cutting heads of roadheaders and cutting drums of longwall shearers are commonly recovered in the overhaul process. Their dimensions often don't match the nominal values. Control of the process of pick holders positioning in robotic technology determines the need to develop a method that allows on–line monitoring and measuring the distance between the pick holder base and the side surface of the cutting head at the place of its assembly, while the robot arm travels to the target position resulting from the position of the given pick holder.

Having information about the spatial location of the cutting head side surface points and the pick holder base points, it is possible to determine the position of the pick holder relative to the body of the cutting head (cutting drum). When the actual shape of this body deviates from the nominal, it will be possible to adjust the position of the pick holder to eliminate the risk of collision of the pick holder with the cutting head body and ensure its welding.

The method presented in this article allows the distance between the pick holder base and the side surface of the cutting head to be measured at any point of the pick holder base. It is based on the analysis of measurement photos obtained from the cameras of the KUKA VisionTech vision system. These photos are processed by the Computer Vision Toolbox in MatLab software. It is also possible to use other tools, such as the OpenCV library [3] or the LabView software.

2. The method of determining the distance between the pick holder base and the side surface of the cutting head using stereoscopic vision

There are many possibilities to obtain a stereoscopic effect [4]. Three dimensional vision is based on it. If a point is recorded on at least two photos taken from different positions, it is possible to clearly determine its coordinates in space [5]. Depending on the possess input data, there are different ways to determine these coordinates [6].

The starting point of the measurement process is the camera or cameras calibration. It is done in order to obtain matrix of camera internal parameters [7]. The camera's internal parameters determine the optical properties of the lens, its radial and tangential distortions. The coordinates of the image center (projection center) are also determined. There are several ways to calibrate cameras, e.g. using a calibration plate. In use are methods such as: Direct Linear Transform, Tsai, Zhang, Heikkil, Bouguet and many more [8]. Calibration of cameras can be done on the basis of one or many photos. Some methods include radial and tangential distortions, and some do not. Regardless of the method used, the results are quite similar. In the developed measuring method the Zheng method was used. This is the default method implemented in the MatLab software, which uses the default calibration plate (Figure 1a). To calibrate the cameras, eleven photos were taken at various positions relative to the calibration plate (Figure 1b).

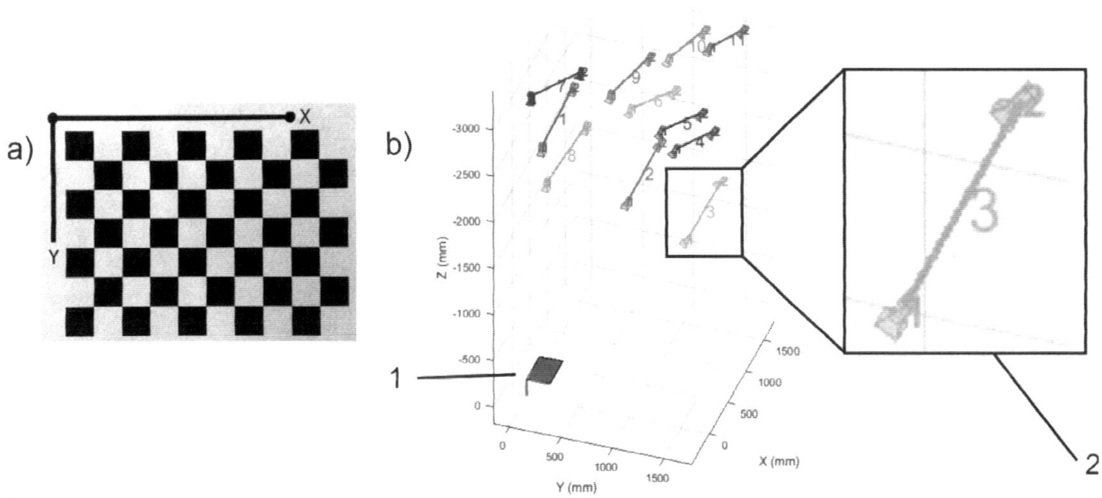

Figure 1. Calibration table (a) and camera position relative to the calibration table during calibration (b): 1 – calibration plate, 2 – camera set (colored segments marked with numbers from 1 to 11 show the position of the camera set base relative to the calibration plate).

To get a stereoscopic effect it is not necessary to use two cameras [9]. Algorithms implemented in computer tools allow 3D visualization of points visible on downloaded images without knowledge of the external parameters of the camera system [10]. The external parameters of the camera system determine the offset and rotation of the cameras relative to the coordinate system associated with the observed scene. Using the appropriate functions and based on certain corresponding points in both images, it is possible to build a spatial model of the photographed scene. Nevertheless, having knowledge about the external parameters of cameras, this reconstruction is easier and more accurate [11, 12]. Different calibration methods require different numbers of measuring points needed for calculations. They are used among others: sevenpoint algorithm, eightpoint algorithm, Ransac, or LMedS algorithm [13]. As a result of their operation we get the values of matrix elements that determine the relationship between the points recorded in the pictures. Camera calibration is enabled by computer tools. The basic method in the MatLab software is the eightpoint algorithm.

With an industrial robot at its disposal, it is possible to acquire subsequent measurement photos with one camera, by changing its position in space using a robot [14]. Then you can get the same effect as using two cameras.

If two cameras are used, knowing their external and internal parameters, using triangulation, you can determine the spatial coordinates of the corresponding points on the pair of analyzed photos [15]. If the camera layout is canonical, then a disparity map can be computed based on the camera's internal and external parameters [16, 17]. It illustrates the difference in pixel position between a pair of stereo images [18]. The spatial model built on the basis of the disparity map includes the scene recorded in the pictures.

In order to determine the distance between the side surface of the cutting head body and the base surface of a given pick holder, it is necessary to determine the shape of the cutting head side surface in the immediate vicinity of the considered pick holder. This is done on the basis of the location of markers applied to the side surface of the cutting head body [19]. During the bench tests, these markers were in the form of grids of dots printed on a sheet of paper (Figure 2a). This card was glued onto the body of the cutting head in the place where the pick holder base will be located after placing it with the robot arm. An alternative and more practical solution is projection of the dot grid with a projector or laser [20] (Figure 2b). Due to the fact that the pick holder base is flat, to determine its position in space it is sufficient to determine the coordinates of three of its four corners.

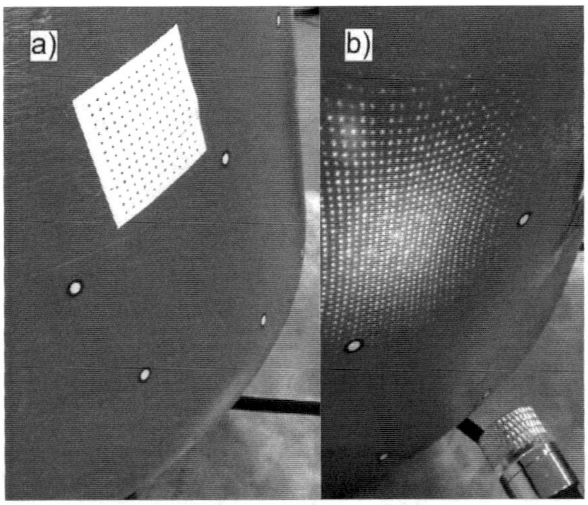

Figure 2. Markers applied to the body of the cutting head: a) in the form of a grid printed on a sheet of paper, b) displayed with a laser.

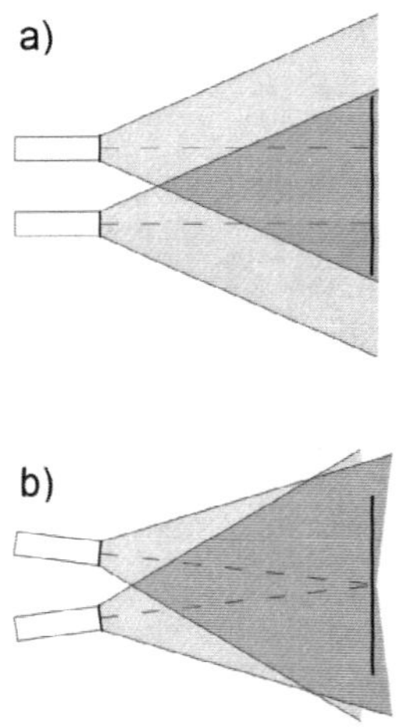

Figure 3. Canonical camera arrangement (a) and converging camera arrangement (b).

Two digital cameras, which are part of the vision system, were used in the bench tests. The use of two cameras speeds up the acquisition of measurement data. Despite the fact that they are the same cameras from the same manufacturer, the photos recorded by these cameras may be of different quality. This makes it difficult for the software to find the corresponding points in the images. Various techniques to improve photo quality come to the rescue, such as: segmentation, contrast equalization methods, line filters, morphological operations, or histogram equalization methods [21].

Due to the mutual arrangement of cameras, they can work in a canonical (parallel) or convergent arrangement [22]. Because two narrow–angle cameras were used in the research and the areas that were photographed are small, a converging camera arrangement was used (Figure 3b). In the case of a canonical system (Figure 3a), these cameras would have to be far away from the photographed objects during image acquisition to ensure an adequate ratio of base length (distance between cameras) to the distance from the photographed objects [23]. When the cameras are close to the photographed object, it is reproduced with a higher resolution. For this reason, markers applied to the cutting head side surface should be as small and densely spaced as possible, which increases measurement accuracy. As a result of applying a convergent camera system, we get two photos of objects of interest to us rotated relative to each other. Similarly to the method described in the work [24], cameras should be placed in an appropriate manner relative to the base of the pick holder and the side surface of the cutting head body during the acquisition of measurement photos.

The measurement using the stereo vision system is carried out during the positioning of the pick holder relative to the side surface of the cutting head body. During the approach with the pick holder to the side surface, measurement photos are acquired. The advance direction of the gripper built on the robot arm, holding the positioned pick holder, is perpendicular to the pick holder base. Because the robot control system uses the position and orientation of the tool coordinate system ($X_TY_TZ_T$) when positioning the pick holders, it must be transformed into the coordinate system associated with the pick holder base ($X_PY_PZ_P$).

Two camera settings were considered in relation to the photographed objects – upper (marked as "U") and bottom (marked as "B") (Figure 4). Photographic stands are located on two opposite sides of a given pick holder in such a way as to ensure that cameras record all points of interest. Axes: X_U and X_B of the coordinate system associated with the vision system for its individual settings coincide with the plane extending between the axis of rotation of the cutting head and the center of the pick holder base (plane 1). If all the points cannot be found on the recorded photos from the upper position, the robot control system positioning the vision system can set the cameras in the bottom position. Determination of the position and orientation of the vision system in space for individual photogrammetric stations is possible based on composite homogeneous transformation matrices:

– upper setting: $$\mathbf{B}_U = \mathbf{Rot}(\boldsymbol{Z}, \boldsymbol{\alpha}_{DP}) \cdot \mathbf{Rot}\left(\boldsymbol{X}, -\frac{\pi}{2}\right) \cdot \mathbf{Rot}(\boldsymbol{Y}, \boldsymbol{\beta}_{DP}) \cdot \mathbf{Trans}(\mathbf{0}, \mathbf{0}, -\boldsymbol{l}) \quad (1)$$

– bottom setting: $$\mathbf{B}_B = \mathbf{Rot}(\boldsymbol{Z}, \boldsymbol{\alpha}_{DP}) \cdot \mathbf{Rot}\left(\boldsymbol{X}, \frac{\pi}{2}\right) \cdot \mathbf{Rot}(\boldsymbol{Y}, \boldsymbol{\beta}_{DP}) \cdot \mathbf{Trans}(\mathbf{0}, \mathbf{0}, -\boldsymbol{l}) \quad (2)$$

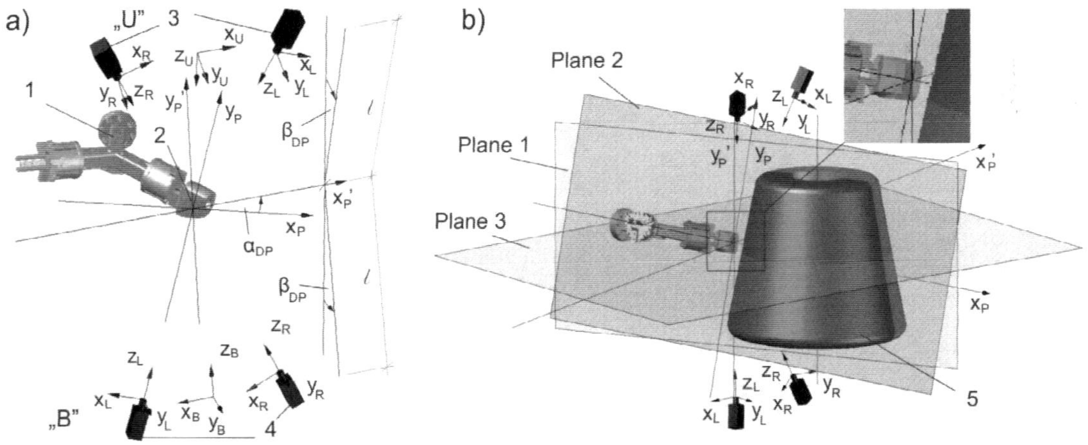

Figure 4. Camera settings during the acquisition of measurement photos: a) relative to the pick holder base, b) relative to the side surface of the cutting head body: 1 – gripper, 2 – pick holder, 3 – cameras in the upper position (located from the front side of the cutting head), 4 – cameras in the bottom position (located from the base of the cutting head), 5 – side surface of the cutting head body.

where:

α_{DP} – angle between the axis X_P coordinate system associated with the base of the given pick holder and axis X_P' coordinate system $X_P'Y_P'Z_P'$, being part of the common plane of the pick holder base (plane 2 in Figure 4b) and the plane perpendicular to the axis of rotation of the cutting head's side surface (plane 3 in Figure 4b),

β_{DP} – angle between the pick holder base and the camera's optical axis,

l – distance of the origin of the vision system coordinate system from the origin of the coordinate system of the pick holder base ($X_PY_PZ_P$).

Because the spatial coordinates of the markers obtained as a result of measuring photos processing in the Matlab software are specified in the coordinate system associated with the left camera ($X_LY_LZ_L$), to obtain the coordinate values of these points in the coordinate system for the base of the pick holder, we must first transform the coordinate system into the system coordinates associated with the vision system ($X_UY_UZ_U$ or $X_BY_BZ_B$) (Figure 5). The camera coordinate system is so oriented that the Z axis coincides with the optical axis of the camera. The X and Y axes form a plane parallel to the background plane (CCD image sensor). To make such a transformation, the position vector of each measuring point should be multiplied by a homogeneous composite transformation matrix C:

$$\mathbf{C} = \mathbf{Rot}(Y, \alpha_U) \cdot \mathbf{Trans}(-l_X, 0, -l_Z) \quad (3)$$

where:

α_U – the angle contained between the optical axis of the cameras and the Z_U axis (or Z_B axis) of the coordinate system associated with the vision system,

l_X – distance of the origin of the coordinate system associated with the left camera optical system from the origin of the coordinate system $X_UY_UZ_U$ (or $X_BY_BZ_B$) cameras along the axis X_U (X_D),

l_Z – distance of origin of the coordinate system $X_UY_UZ_U$ (or $X_BY_BZ_B$) from the origin of the left camera coordinate system measured along the axis Z_U (Z_B).

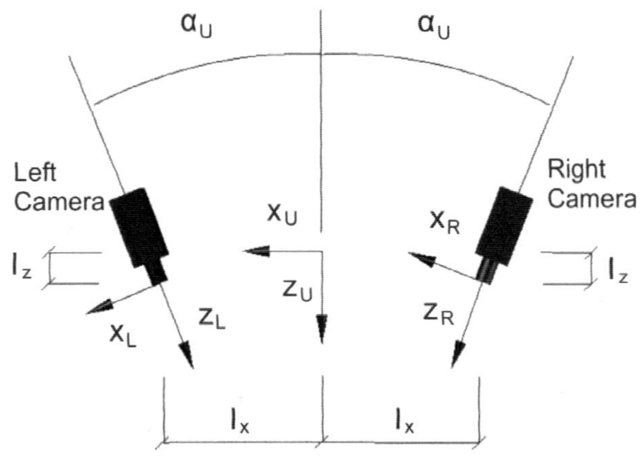

Figure 5. Relations between coordinate systems associated with cameras ($X_LY_LZ_L$ and $X_RY_RZ_R$) and the coordinate system of the vision system in the upper position ($X_UY_UZ_U$).

In the second stage, the vision system coordinate system ($X_UY_UZ_U$ or $X_BY_BZ_B$) is transformed into the coordinate system of the pick holder base ($X_PY_PZ_P$). It is described by the following composite homogeneous transformation matrices:

– for upper setting of the vision system:

$$\mathbf{D}_U = \mathbf{Trans_1}(0, 0, l) \cdot \mathbf{Rot_1}(Y, -\beta_{DP}) \cdot \mathbf{Rot_1}\left(X, \frac{\pi}{2}\right) \cdot \mathbf{Rot_1}(Z, -\alpha_{DP}) \quad (4)$$

– for bottom setting of the vision system:

$$\mathbf{D}_B = \mathbf{Trans_2}(0, 0, l) \cdot \mathbf{Rot_2}(Y, -\beta_{DP}) \cdot \mathbf{Rot_2}\left(X, -\frac{\pi}{2}\right) \cdot \mathbf{Rot_2}(Z, -\alpha_{DP}) \quad (5)$$

As a result, the components of the position vectors of each measuring point are determined in the coordinate system associated with the pick holder base ($X_PY_PZ_P$):

$$[x_{Pi,}, y_{Pi}, z_{Pi}, 1]^T = \begin{cases} C \cdot \mathbf{D}_G \cdot [x_{Li,}, y_{Li}, z_{Li}, 1]^T \\ C \cdot \mathbf{D}_D \cdot [x_{Li,}, y_{Li}, z_{Li}, 1]^T \end{cases} \quad \text{for } \boldsymbol{i} = \mathbf{1}, \ldots, \mathbf{N}, \tag{6}$$

where the coordinate z_P is the distance measured in the advance direction of the gripper's during the positioning of the pick holder (Figure 6). All points that are found within the pick holder base are considered, i.e. those whose x_P and y_P coordinate values are within the rectangle representing the pick holder base.

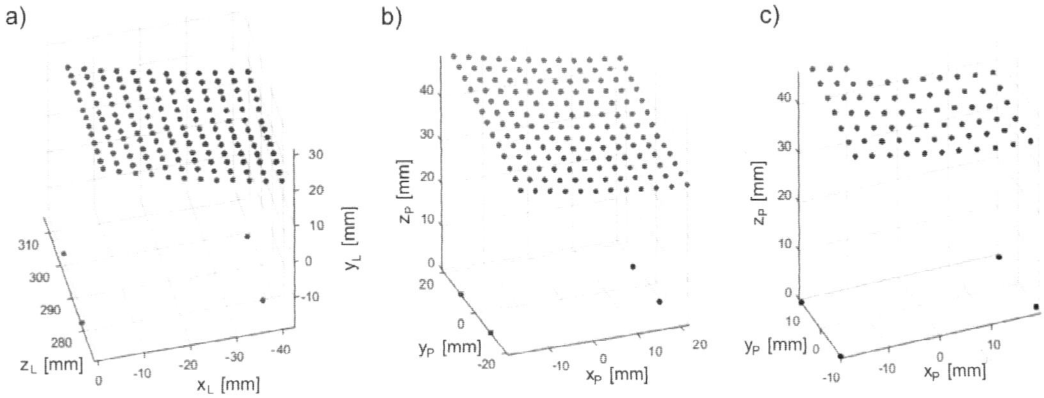

Figure 6. Location of markers: a) in the coordinate system associated with the left camera $X_LY_LZ_L$, b) in the coordinate system related to the pick holder base $X_PY_PZ_P$, c) located within the pick holder base in the coordinate system associated with the pick holder base.

If there no data on the position of the vision system during the acquisition of measurement photos, we can specify the values of the marker coordinates in the coordinate system associated with the base of the pick holder based on information about the location of the characteristic points of this base. For this purpose, the coordinate system associated with the left camera ($X_LY_LZ_L$), in which the coordinate values of the measuring points are defined, should be transformed into the coordinate system associated with the pick holder base ($X_PY_PZ_P$). It comes down to the translation of the camera coordinate system to the point of the pick holder base with known coordinates (e.g. one of the corners). The components of the translation vector are equal to the coordinates of this point. Then we rotate this coordinate system so that the axes of the X and Y system are parallel to the edge of the pick holder base.

3. Measuring stand

The developed measurement method was tested in the Laboratory of robotics of the Department of Mining Mechanization and Robotisation of the Faculty of Mining, Safety Engineering and Industrial Automation at the Silesian University of Technology (Figure 7). Pick holders taken from the storage are positioned using a KUKA KR 16–2 robot (1) located on the KL 250–3 linear unit (2), relative to the side surface of the cutting head (4) placed on the rotary table of the positioner PEV–1–2500 (3). For the purposes of testing, the cameras (9) were placed on a tripod with a special design bracket (10) attached to it – Figure 7b. KUKA MXG20 cameras (9) are connected to the robot control cabinet (6) with a switch (8). Measurement photos are taken from the robot's memory and processed on a computer with MatLab software installed along with appropriate toolboxes. In the developed measurement program, pairs of photos are processed, resulting in the spatial coordinates of the points representing the pick holder base and the side surface of the cutting head body within its place of assembly.

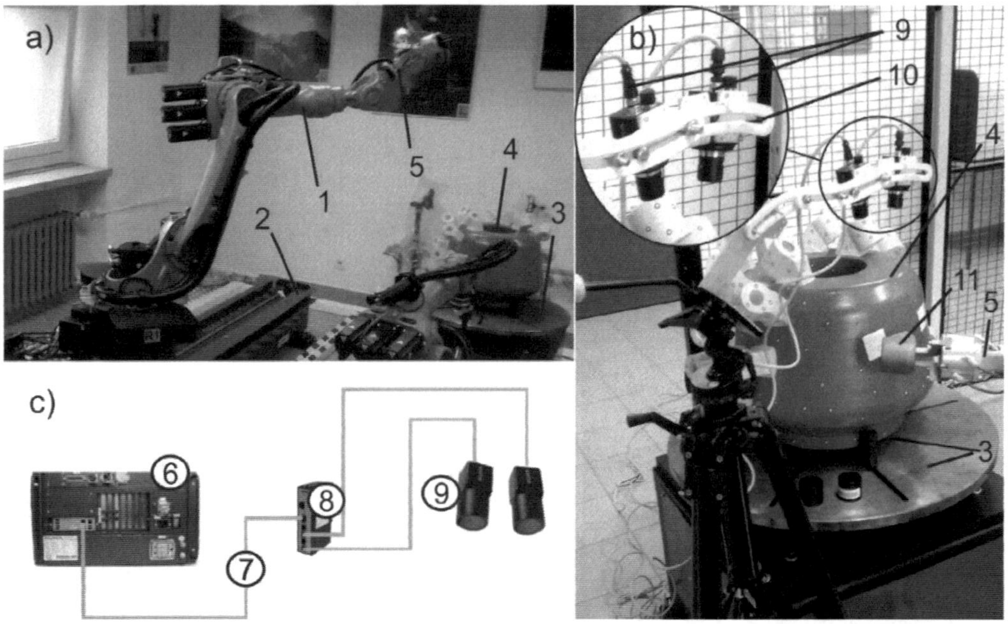

Figure 7. Experimental stand in the Laboratory of robotics of the Department of Mining Mechanization and Robotisation (a & b) and a schematic diagram of cameras connection to the robot control system (c): 1 – KUKA KR 16–2 robot, 2 – KL 250–3 linear unit, 3 – PEV – 1–2500 positioner, 4 –cutting head body, 5 – gripper, 6 – robot controller, 7 – cable, 8 – switch, 9 – two cameras, 10 – special bracket, 11 – positioned pick holder.

The set of spatial coordinates of markers located on the side surface of the cutting head body is obtained as a result of processing and analyzing measurement photos. For this purpose, functions and procedures implemented in the measurement program from the Computer Vision Toolbox are used. Each of these functions is responsible for a certain process. The *imadjust* function adjusts the intensity of the raw image (Figure 8), which changes the contrast so that the pixels clearly stand out in the image (Figure 9). Based on the camera calibration data, the *undistortimage* function removes distortions from downloaded images, resulting from optical distortions. Corresponding measuring points in the pictures are adjusted using the *cpselect* function. Basis on the position of the measuring points in the pictures and the values of external camera parameters, the *triangulate* function determines the values of the spatial coordinates of the measuring points in the coordinate system associated with the left camera ($X_LY_LZ_L$). Then, using the formulas (3) – (6), the coordinate values of these points are determined in the $X_PY_PZ_P$ coordinate system.

The developed method for measuring the position of the pick holders on the side surface of the cutting head body is based on the identification of measuring points (markers). The higher the density of points marked on a given area of the side surface of the cutting head body during acquisition of measurement photos, the more accurate its 3D model will be. Having the coordinates of the side surface points in the coordinate system associated with the pick holder base, we can determine the distance distribution between this base and the side surface of the cutting head body in the place where the pick holder is to be mounted. The distance between the pick holder base and the side surface between the measuring points can be determined using linear interpolation. The algorithm of the measurement procedure is shown in the Figure 10.

4. Results of bench tests

Using the proposed method, the distance between the pick holders base and the side surface of the cutting head body was measured. Figure 11 shows examples of measurement results. The position and orientation of the coordinate system associated with the robot's positioning tool considered pick holder during the measurement, describe the values of the gripper's TCP point: x_T = 1088 mm, y_T =

1437 mm, z_T = 951 mm and angles defining the orientation of the robot tool coordinate system in the basic system: A = 47 deg, B = 10 deg, C = 159 deg. Following the algorithm of the measurement procedure (Figure 10), the distribution of the distance between the surface of the pick holder base and the side surface of the cutting head body was determined (Figure 11). These distances correspond to the coordinate values z_P.

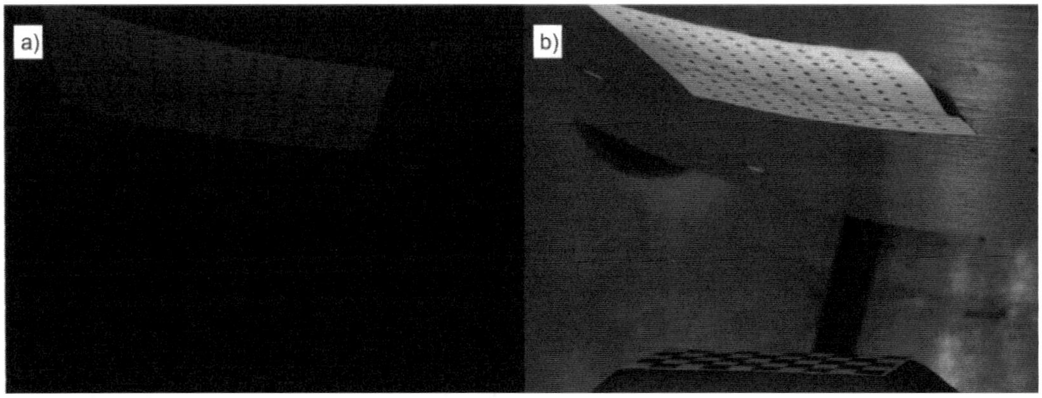

Figure 8. Raw photos taken during measurements: a) image from the left camera, b) image from the right camera.

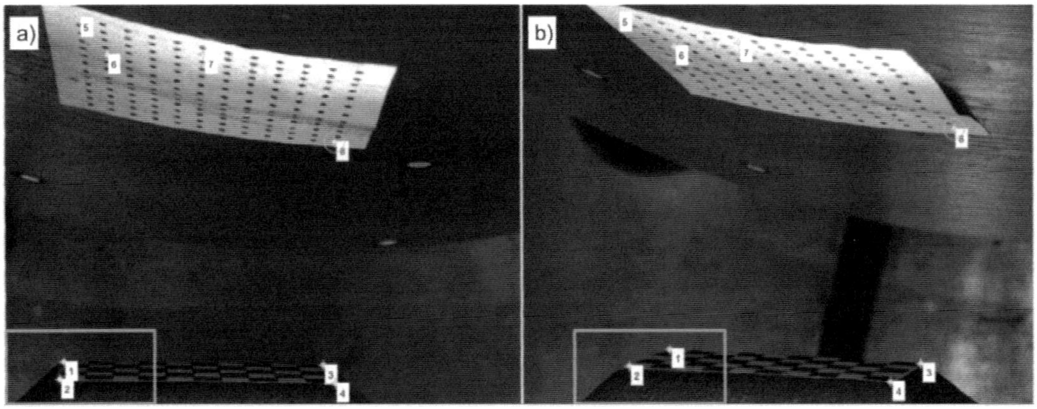

Figure 9. Application of the *imadjust* function – corresponding points are visible in the pictures: a) from the left camera, b) from the right camera.

The shortest distance between the surfaces considered is sought for. It determines how much the pick holder should be moved to the side surface of the cutting head body. This distance should correspond to the distance that the robot tool has to travel to set the pick holder according to the project design. In the case under consideration, the shortest distance is 35.1 mm. Due to the shape of the side surface of the cutting head body and the way the pick holder is positioned, the closest points are located within one of the corners.

5. Summary

The developed measuring method allows real–time determination of the distance distribution between the pick holder base and the side surface of the cutting head/drum body. This is particularly important in the case of re–use in the manufacturing process of the bodies of cutting heads/drums from which old pick holders have been removed. In this case, because the side surface of this body, if it is not subjected to regeneration aimed at restoring its nominal shape, is uneven. This is due to the remains of weld residues connecting removed pick holders and the defects created during their removal.

IOP Conf. Series: Materials Science and Engineering **679** (2019) 012005 doi:10.1088/1757-899X/679/1/012005

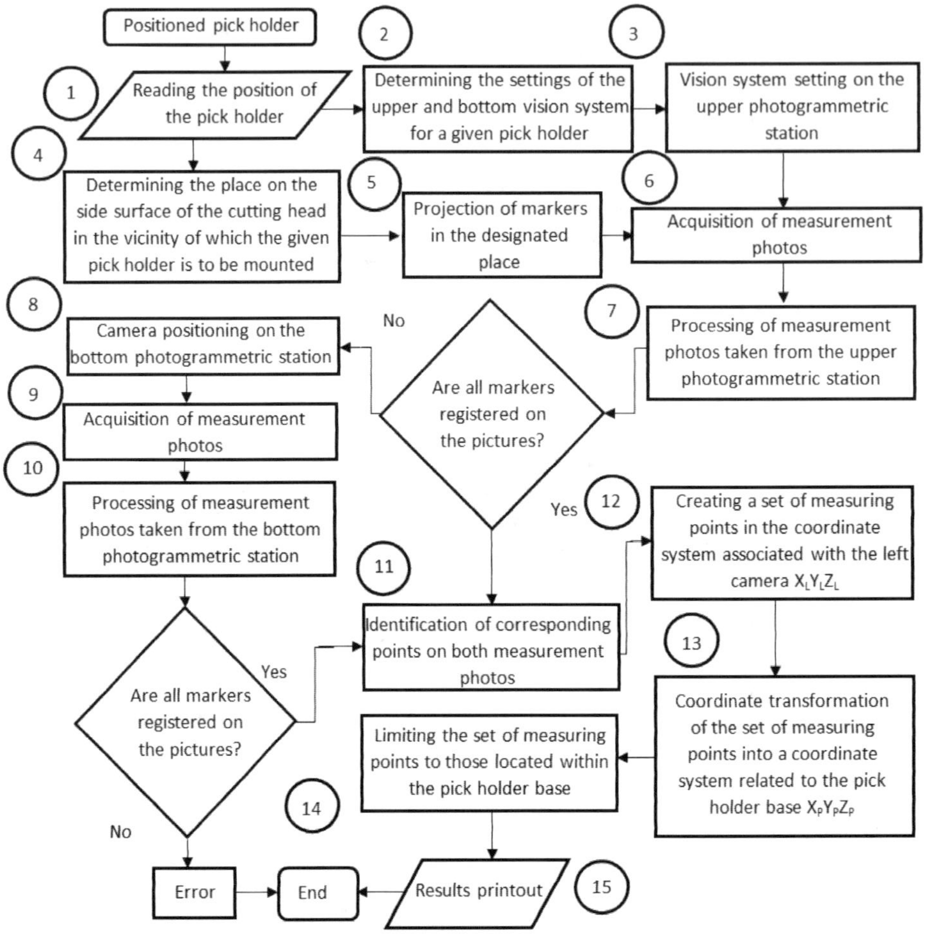

Figure 10. Algorithm of the measurement procedure.

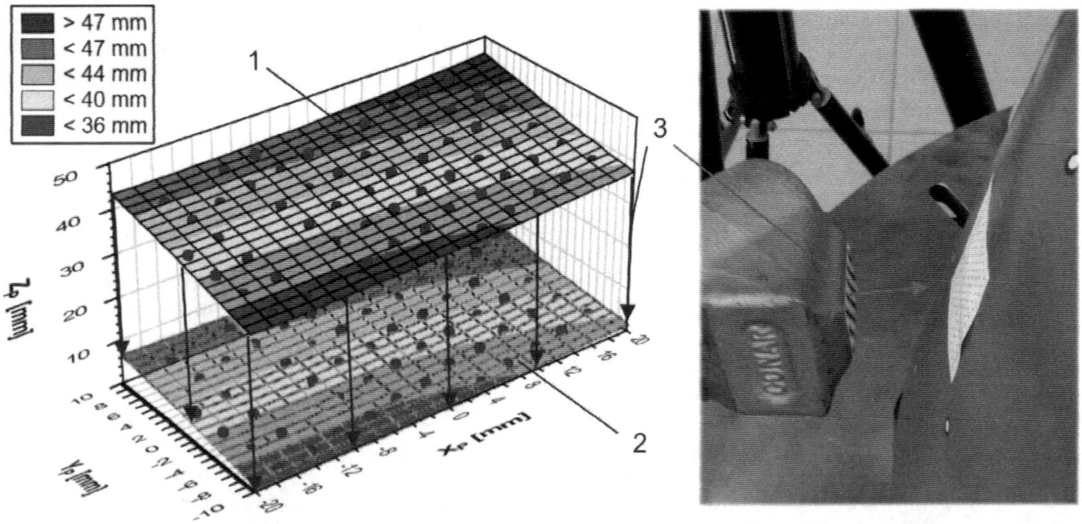

Figure 11. Distribution of the distance between the pick holder base and the side surface of the cutting head body: 1 – measured values, 2 – values after moving the pick holder to the side surface of the cutting head body, 3 – direction in which the pick holder is moved to the side surface of the cutting head body (the direction of the robot tool advance).

In robotic technology for manufacturing cutting heads/drums, such a situation may lead to collision of the pick holders with the side surface of their body or prevent the assembly process (welding).

On–line measurement of the distance of the base of individual pick holders from the side surface of the cutting head/drum during robotic positioning lies at the basis of the adaptive control robot positioning the pick holders. It is possible on this basis to adjust the pick holder position so as to prevent the abovementioned situations during the manufacturing of these important, from the point of view of efficiency of the rock mining process, elements of mining machines.

Acknowledgments
The publication is supported by the Silesian University of Technology's own scholarship fund, in 2018 – 2019.

References

[1] Cheluszka P 2010 Automated assembly of pick boxes placed on cutting heads of winning machines *Mechanizacja i Automatyzacja Górnictwa* **11(477)** pp 11–17 (in Polish)
[2] Dolipski M, Cheluszka P and Sobota P 2008 *Robotic station for positioning pick holders on the side surface, especially heads and drums mining machines and the manner of setting pick holders* patent description 216341 [2010, 07, 386580] (in Polish)
[3] Omosekieji G 2018 *Industrial vision robot with Raspberry Pi using PIXY camera Information* Technology Embedded System Engineering
[4] Schreer O, Brandenburg N and Kauff P 2000 *A Comparative Study on Disparity Analysis Based on Convergent and Rectified Views* BMVC2000
[5] Graefe V, Wershofen K P and Huber J 1993 Dynamic Vision for Precise Depth Measurement and Robot Control *Computer Vision for Industry München*
[6] Peyman A 2015 *Object Distance Measurement Using a Single Camera for Robotic Applications*
[7] Nedevschi S, Marita T, Vaida M, Danescu R, Frentiu D, Oniga F, Pocol C and Moga D C*amera calibration method for stereo measurement*
[8] Szeliski R 2010 *Computer Vision: Algorithms and Applications* Springer
[9] Wubbena H, Nieves E, Nawab A and Garmann G 2011 *Welcome to Accurate Robotic 3D Vision* An educational webinar sponsored by Universal Robotics and Yaskawa Motoman Robotics
[10] Fusiello A and Irsara L 2011 Quasi–Euclidean Epipolar Rectification of Uncalibrated Images *Machine Vision and Applications* **22** Issue 4 pp 663–670
[11] Kyto M, Nuutinen M and Oittinen P 2011 *Method for measuring stereo camera depth accuracy based on stereoscopic vision* The International Society for Optical Engineering
[12] Takahachi G and Matsuoka R 2010 *Accuracy of measurement using a pair of stereo images acquired by finepix real 3D w1 without controls,* International Archives of Photogrammetry, Remote Sensing and Spatial Information Sciences (Newcastle upon Tyne UK) Vol XXXVIII Part 5
[13] Hartley R and Zissermann A 2003 *Multiple View Geometry in Computer Vision* Cambridge University Press
[14] Ersu E and Wienand S 1995 *Method for determining the position of a body in space* european patent EP 0 763 406 A1 [C.A., 1997, 19, 95114588.7] (in German)
[15] Rzeszotarski D, Strumiło P, Pełczynski P, Wiecek B and Lorenc A 2005 Stereovision system for 3D reconstruction of image sequneces *Elektronika: prace naukowe* **10** pp 165–184 (in Polish)
[16] Cutolo P and Ferrari V 2018 The Role of Camera Convergence in Stereoscopic Video See–through Augmented Reality Displays *International Journal of Advanced Computer Science and Applications* **9** No 8
[17] Wright S 2011 *Parallel vs Converged the dabate over stereo camera*
[18] Fusiello A, Trucco E and Verri A 2000 A compact algorithm for rectification of stereo pairs *Machine Vision and Applications* **12** Issue 1 pp 16–22
[19] Jeżewski S and Jaros M 2008 Scanning of three–dimensional room space *Automatyka* **12** no 3 pp 669–673 (in Polish)

[20] Makoto Iwamoto 2018 Active pattern projection improves AOI 3D measurement accuracy *Vision Systems Design, vision and automation solution for engineers and integrators worldwide* **23** no 2 pp 7–10

[21] Keselman Y and Dickinson S 2001 *Bridging the Representation Gap Between Models and Exemplars* In IEEE Conf. on Comp. Soc. Work. on Models versus Exemplars in Comp. Vis

[22] Bernasik J and Mikrut S 2007 *Engineering photogrammetry* (Kraków) (in Polish)

[23] Gallup M, Frahm J M, Mordohai P and Pollefeys M 2008 *Variable Baseline/Resolution Stereo* CVPR08

[24] Cheluszka P and Jagieła–Zając A 2019 Determining the position of pick holders on the side surface of the working unit of the cutting machine in the robotic technology of their assembly *IOP Conference Series: Earth and Environmental Science* Volume **261** Number 012003

IOP Conf. Series: Materials Science and Engineering **679** (2019) 012006 doi:10.1088/1757-899X/679/1/012006

Implementation of a jig control system at BUDRYK Coal Mine

S Jendrysik[1]*, D Jasiulek[1], K Stankiewicz and J Babczyński[2]

[1] KOMAG Institute of Mining Technology, Division of Mechatronic Systems, Pszczyńska 37, 44-101 Gliwice, Poland
[2] Jastrzębska Spółka Węglowa S.A. KWK "Budryk" Zamkowa 10, 43-178 Ornontowice, Poland

*E-mail: sjendrysik@komag.eu

Abstract. The article presents conceptual assumptions and achieved functional-utility effects concerning the implementation of control and visualization system of operation of pulsating jigs in the Coal Mechanical Processing Plant of BUDRYK Coal Mine. The system includes six OSM medium-grain and two OM fine jigs. The communication structure of the control system, including the connection with the master controller, is presented. The process of selecting the controller settings, to stabilize the quality parameters of the coal beneficiation, is discussed.

1. Introduction

Pulsating jigs are commonly used in processing plants of Polish mines. This popularity is related, among others, to their reliability and a wide range of grain classes of the feed sent for beneficiation [1]. Coal beneficiation in a jig, which consists in among others, material loosening and transporting, is a complex process, characterized by multidimensionality, non-linearity, fluctuation of parameters and nonstationary character of disturbances. Increasing level of requirements in relation to the quality of coal concentrate, obtained at the jig output, requires to use modern, safer and more reliable control systems. Solutions from the broadly understood field of industrial automation, using modern microprocessor systems, efficient and fast communication modules, come to support. Their implementation enables comprehensive control over the entire production process, its selected fragments or of a single machine.

As a part of development and modernization of the Mechanical Coal Processing Plant at BUDRYK Coal Mine, 8 new jigs, equipped with modern KOGASTER control systems, are implemented. At the time of writing this article, three jigs were put into operation, the remaining ones are under construction.

2. Structure of the jig beneficiation system in BUDRYK Coal Mine

The system installed in the BUDRYK Coal Mine includes 8 jigs, which form three beneficiation systems (Figure 1):

- beneficiation of coal type 34 - 3 OSM jigs,
- beneficiation of coal type 35 - 3 OSM jigs,
- beneficiation of secondary coal - 2 OM jigs.

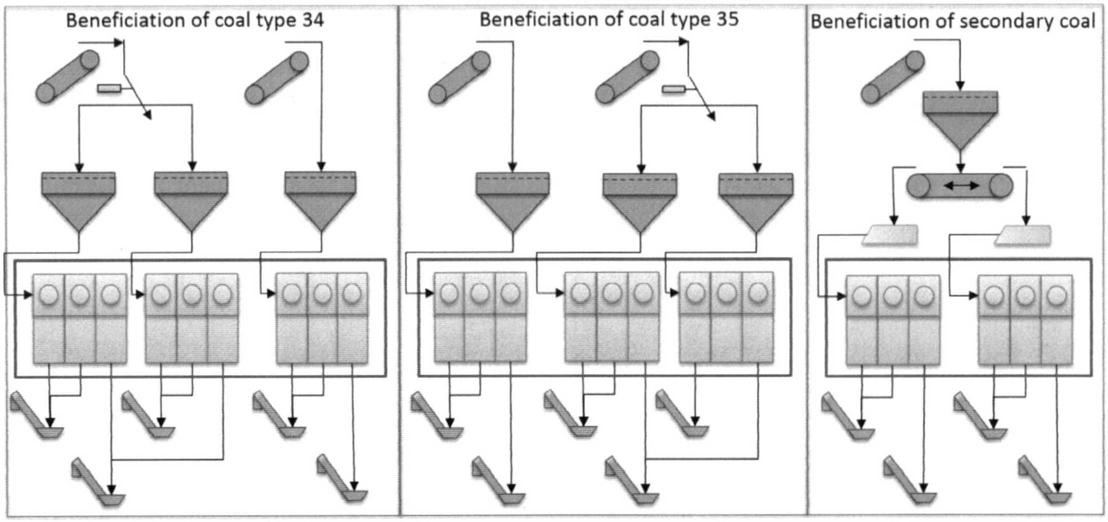

Figure 1. Simplified structure of beneficiation systems in BUDRYK Coal Mine.

Each jig is divided into three sections separating the beneficiation products. In the first and second sections the waste product is obtained, in the third section an intermediate product and a concentrate are received. The facility is supplied with coal 34 and 35 by 4 belt conveyors B-1200 (two from each buffer tank). The excess material from the OSM jigs is transported to the secondary beneficiation node. Heavy products from the jigs are received by the B-1000 bucket conveyors. These conveyors are equipped with frequency converters, which allow to adjust their efficiency to the current load [2].

3. Characteristics of the jig control system

A correct realization of beneficiation process, taking place in the jig, requires use of an appropriate control system for devices being a part of the node. In the process of hard coal beneficiation, taking place in pulsating jigs, a very important element influencing its proper progress, has an efficient and precise control over raw material feed and output of enriched product [3, 4]. It is related to the need of a complex node automation, which can provide proper interaction of jig with the feed system and bucket conveyors, receiving products from the jig. The combination of controllers of above mentioned machines, the use of distributed input-output modules to operate the executive elements and measuring devices, the use of operator panels and visualization-archiving systems and ensuring reliable data transmission among the automation elements, form an integrated system ensuring comprehensive control of the entire process.

The jigs control system at the BUDRYK Coal Mine is divided into 4 control cabinets. One cabinet supports 2 jigs. In each cabinet there is an independent PLC installed to supervise an operation of two machines. In each of the 3 sections of a jig the following parameters are measured (Figure 2):

- height of the separation layer,
- opening degree of the heavy product bleed gap,
- pressure in the air manifolds,
- pressure in the equalisation reservoirs,
- lower water flow rate.

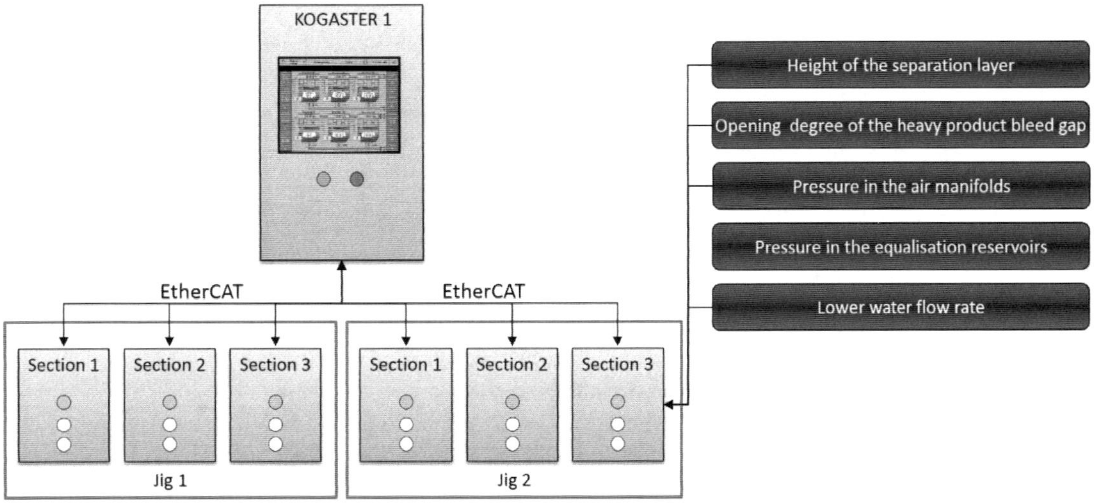

Figure 2. Structure of the measuring system.

In addition the following parameters are also measured:

- feed flow rate,
- control air pressure,
- pressure in the hydraulic system,
- load on bucket conveyor motors.

The following devices are also connected to the control cabinet of the jig control system:

- compressor for preparing control air,
- hydraulic aggregate supplying cylinders of heavy products reception systems,
- throttles of lower water supplied to jig,
- throttles of working air.

Controlled parameters are:

- draining flow rate of the heavy product,
- lower water flow rate,
- air pressure in the pulsating chambers,
- speed of the bucket conveyor.

The jig control system strictly cooperates with the master system of the processing plant, which manages the operation of machines supplying the feed to a jig, as well as realizing technological interlocks used to stop the whole or a part of node safely. Figure 3 shows the communication structure among the jigs control systems and the master system.

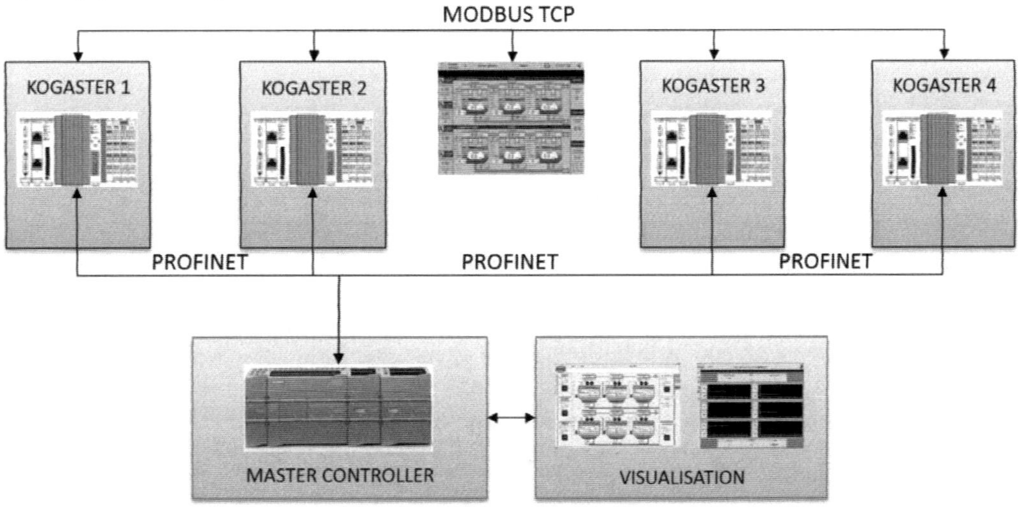

Figure 3. Communication structure of the control system.

The communication was carried out using the Profinet network, a universal communication network based on the industrial Ethernet standard. The internal controller of the jigs control system is connected via ModbusTCP. In addition, using this communication standard, the embedded controller will be connected to an additional visualization system, which will be installed in the soundproof operator's cabin enabling a supervision of the machinery operation.

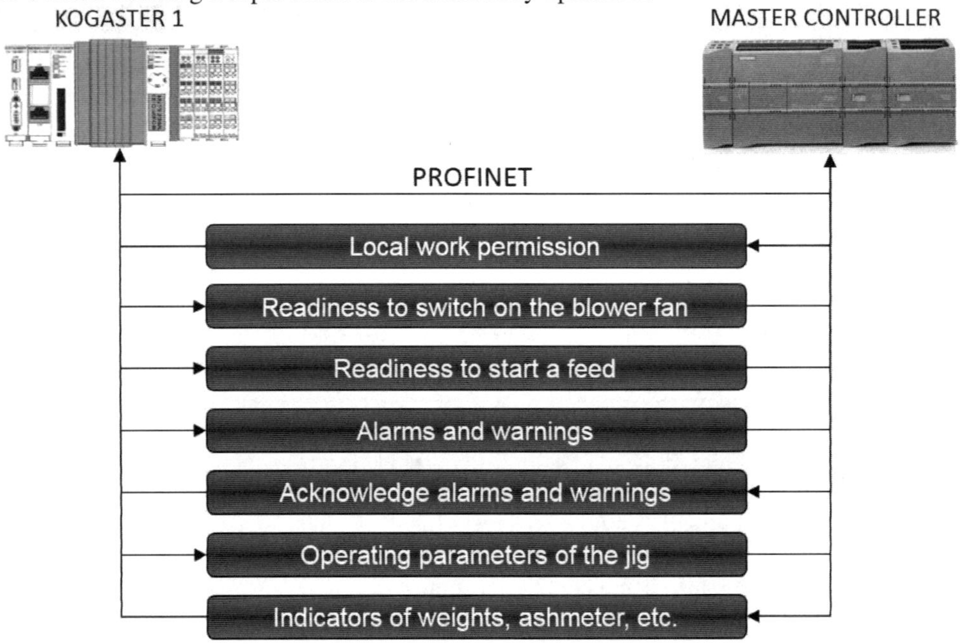

Figure 4. Structure of data exchange.

The diagram showing the procedure of interaction between the master system and the jig control system is shown in Figure 4. A command to switch on the jig is sent from the master system to the jig controller, which first activates the pulsation valves, then it sets the throttle of working air, mounted at the jig input. After the above operations, the information is sent to the master system about the readiness of the jig to switch on the working air blower. After the blower is switched on, the air pressure is checked both in the equalisation tanks and in the pulsation chambers of the jig. At the same time, the lower water valve is opened and the water flow rate at the inlet to each compartment is checked. Correct values of water pressure and flow rate are the basis for the acceptance of the feed. During the jig operation, other parameters of the machine are checked, but the most important of them include correct functioning of the heavy product receiving systems. A detection of a malfunction results in immediate delete of the feed flow, and an activation of the light and sound signal. The feed can also be switched off at the request of the operator, which is done by using a dedicated button located on the control panel.

During the beneficiation process, the controller performs the following functions [5]:

- an operation of actuators, including:
  - switching of the hydraulic pump, supplying the circuits of hydraulic valves, with fault indication when there is no operating signal and pressure drop,
  - a control of float position, with indication of irregularities resulting from incorrect pulsation, or from float sinking, when movement does not take place despite correct pressure pulsation in the pulsating chamber,
  - a control of air pressure, including amplitudes of change, related to float movement and pressure values in the subsequent phases of the cycle, indicating the level of water in the chamber, with an appropriate alarm,
  - a control and signalling of parameters, including bottom water flow rate, oil pressure in hydraulic cylinders, valve control air pressure, feed rate and status of fuse and power supply units in the control cabinet,
- a pulsation control, which includes:

  - switching of inlet and outlet air valves to/from the pulse chambers, according to a pre-set time cycle with a specified cycle length, pause phases, emptying and filling,
  - a stabilization of the average air pressure in the inlet manifold to the pulsation valves through the bypass valve, which improves the working conditions of the blower by reducing the flow oscillation and allows to switch off the pulsation in the case of a short-term supply failure (without switching off the blower),
  - an automatic correction of the inlet and outlet phase and/or a correction of setpoint value of the stabilised air pressure in the inlet manifold, changing the amount of injected air and removed in a single pulsation cycle, performed in the case of an abnormal amplitude of float pulsation, with acceptable limits of changes,
  - an automatic correction of the inlet/outlet phase length ratio in the event of an abnormal average water level in the pulsating chamber indicated by the pressure gauges in order to change the retention air volume in the chambers,
- a control of the bleed gap, which includes:
  - a control loop, step-by-step change of the position of the culvert in order to maintain a pre-set position of the float (the lowest or average position during the pulsation cycle), usually associated with the position of the trigger threshold,
  - an automatic correction of a pre-set float position as a function of simulated substitutional density changes introduced by the operator, which, to a limited extent, replace the current change in weights levelling the float,
  - an automatic correction of the measured position of the float, as a function of the measured degree of the culvert opening, flow of the lower water and the float depth in relation to the drain, taking into account the effects of hydrodynamic and inertial forces affecting the float movement,
  - an automatic correction of simulated float density in the event of a deviation between the set and measured ash content of the concentrate or its calorific value,
  - flushing, aimed at removing grains blocking the flow and making it impossible to close the culvert, performed in the case of closing faults, or involving multiple quick opening and closing of the culvert, with a repeated attempt at increased closing below the previous position and signalling of failure in the case of the test failure,
- a shutdown sequence of the installation, at the operator's request,
- a sequence of switching on the installation, at the operator's request.

4. Adjustment of the heavy product collection system in the BUDRYK Mine conditions

Raw material (feed), delivered to the jig starts to pulsate. This movement is achieved by injecting and releasing air into the pulsation chambers. A separation of feed in a jig is a complex process. A segregation is associated not only with the difference in specific weights, but also with the size of grains, from which larger grains tend to fall down, an amplitude of movement and accelerations depending on the frequency or even the form of pulsation, the flow rate of lower water and the thickness of the material layer on the screen. In the end, the grains with the highest falling speed are located on the sieve deck (heavy product), while the grains with the lowest falling speed are carried to the surface of the material to be beneficiated (light product). One of the basic functions, performed by the machine control system, is a collection of heavy products. The effectiveness of the beneficiation process depends on the location of the separation layer. The separation layer is a layer from which half of the grains are transferred to the light product and half to the heavy product. In the jigs installed in the Budryk Mine, float sensors moving vertically in a mixture of water and grains, according to the frequency of pulsation, were used to measure the height of the grains forming the separation layer. These sensors are mounted directly to the heavy product collection system. A diagram of the bottom product collection system from the Budryk Mine jigs is shown in Figure 5.

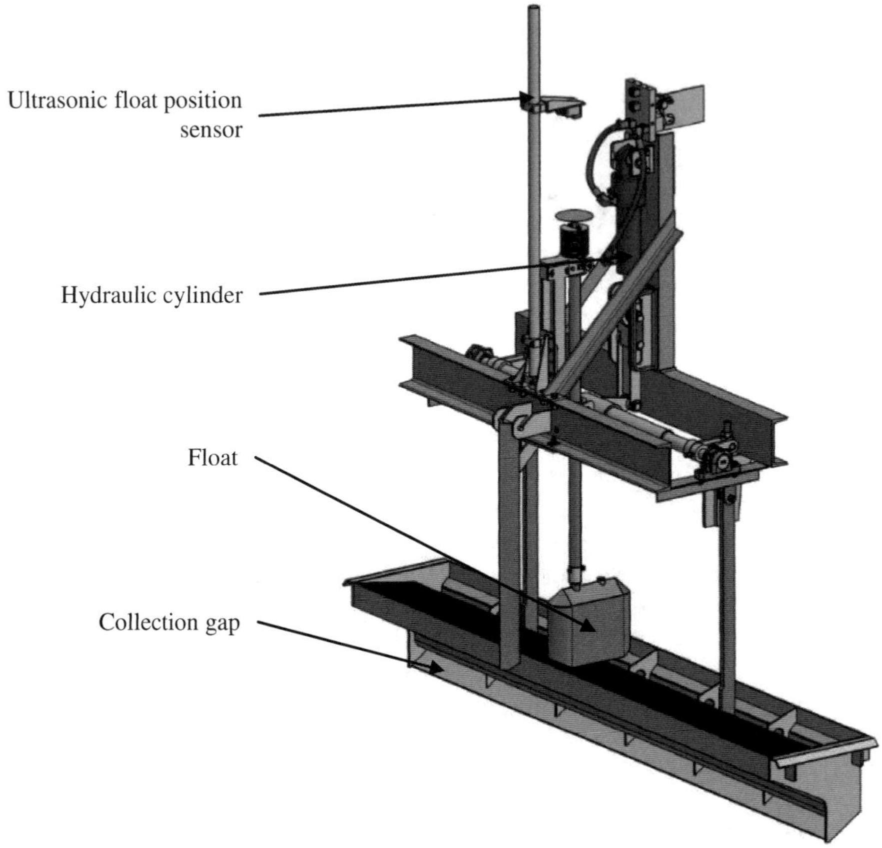

Figure 5. Culvert system in KOMAG type jigs [6, 7].

A collection gap is located at the end of the chamber, before the overflow door. The float ends with a plate, which is a reflective element of the ultrasonic sensor, measuring the float position. On the basis of information from the sensor, the opening of the collection gap is increased/reduced and receiving intensity of heavy product is changed, which results in a change in location of the separation layer (density of separation). The aim is to maintain the separation layer at the level of the overflow door. Disturbances, in form of a change in the quantity and densiometric composition of the feed, cause fluctuations in the separation density. The task of control system is to minimize them. It should be noted that the dynamic parameters of collecting system change under influence of changes in the feed characteristics and are also different for changes in the flow rate of the heavy product. A block diagram of the automatic control system for heavy product collection from the jig is shown in Figure 6.

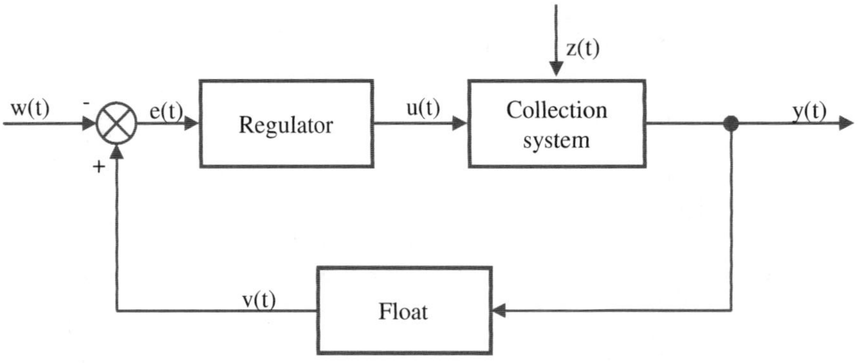

Figure 6. Block diagram of the automatic control system.

The controlled object is the collecting system, which is impacted by the disturbances z(t). This system is equipped with a hydraulic cylinder, which makes it possible to change the opening of the

IOP Conf. Series: Materials Science and Engineering **679** (2019) 012006 doi:10.1088/1757-899X/679/1/012006

bleed gap, thus changing the intensity of the heavy product collection. The measuring device is a metal float, of a specific density, together with an ultrasonic sensor. The sensor sends a feedback signal v(t). This signal is compared with the set value of the separation layer w(t) height, corresponding to the set density of the separation. The e(t) signal received in the comparing system, called the deviation, is transmitted to the regulator, which converts it into a u(t) control signal, which is the basis for changing the opening of the bleed gap (the collection intensity of the heavy product).

The main objective of the regulation is to compensate the impact of disturbing parameters on the regulated ones by changing the setting values so that the adjusted values differ slightly from the required (preset) values.

The most popular PID type regulator is used for this task. In the case of controlling the collection of the heavy product it is essential to optimize the dynamics of the control system in order to eliminate the impact of disturbance and maintain zero error from the steady state extortion. Because of these interferences, which could lead to unpredictable operation of the differential part, from the group of PID regulators, the PI controller with the structure described by the equation was selected:

$$G_{PI}(s) = K_p(1 + \frac{1}{T_i s}) \tag{1}$$

where:

K_p – proportional gain,

T_i – integration time.

The first implementation stage of new jigs in the BUDRYK Mine involved the start-up of two OSM jigs, which are parts of the coal type 35 enrichment system and one secondary beneficiation jig. The OSM jigs are supplied with feed from one belt conveyor. During the commissioning activities, a test was conducted in which one OSM jig was tuned using the Zigler-Nichols method directly on the real object. The other jig was started at the control parameters obtained from simulation tests, using the Matlab/Simulink software.

For simulation tests it was assumed that the jig is characterized by dynamic properties of the first order inertial object, with a delay [8]. The transition function of such a system in the operator form can be written as:

$$Y(s) = \frac{ke^{-s\tau}}{sT + 1} U(s) \tag{2}$$

where:

$Y(s)$ – position of the float (density of separation),

$U(s)$ – opening of the bleeder flap (flow rate of the heavy product),

k – amplification (0.005 for positive changes, 0.003 for negative changes was assumed),

τ – time delay (9 s for positive changes, 28 s for negative changes),

T – time constant (18 s for positive changes, 36 s for negative changes).

The presented model is described with different parameters (k, T, τ) because the dynamic characteristics of the receiving zone in the jig are different for positive and negative changes in the flow rate [8]. For the purpose of the simulation, the current parameters of the OSM jig implemented in the BUDRYK Mine, were assumed as follows:

- output: 250 Mg/h,
- active working surface: 24 m2,
- trough width: 3000 mm,
- trough length: 8000 mm,
- number of beneficiation sections: 3.

It was assumed that the bed had the shape of a cube with dimensions corresponding to the width of the jig and the length of a single section. The thickness of layers was determined from the volume balance of material flowing in and out of the jig.

For simulation tests, the feed characteristics presented in Table 1 was assumed.

Table 1. The feed characteristics used for the regulator initial settings.

Fraction density, g/cm^3	Feed, %
< 1.3	23.78
1.3 ÷ 1.4	9.84
1.4 ÷ 1.5	6.38
1.5 ÷ 1.6	8.81
1.6 ÷ 1.7	7.13
1.7 ÷ 1.8	4.06
1.8 ÷ 2.0	4.29
> 2.0	35.71
Sum	100.00

In the first and second sections, the material of a density > 1.8 g/cm3 was separated, while in the third section the separation density was 1.5 g/cm3. It was also assumed that the movement velocity of individual layers in the horizontal layer changed linearly. The highest velocity was observed in the layer of the lowest density.

The basic task of the regulation system is to adjust the opening of the bleed gap (heavy material flow rate), to increase the opening when the float goes up and to reduce the opening when the float goes down. Tuning the regulator requires an implementation of such solutions that the regulation system acts in accordance with the requirements and meets the adopted criteria of regulation quality. In the real system, constantly exposed to disturbances, statistical parameters of the error signal are analyzed. The first accepted measure of the control quality was the absolute mean value of the error signal. It is required that in the steady-state of the system, this value shall be as low as possible:

$$\min|\bar{e}| = \frac{1}{n-1}\sum_{i=0}^{n} e(i) \tag{3}$$

where:
$\bar{e}$ – average value of the control error
e(i) – discrete series of values representing an error in the control signal.

When tuning the controller, it is necessary to take into account the operation of the system in transition states. The ITSE (Integral Time Square Error) indicator is a good measure of the control quality used in fixed-value systems, where the transition process is to be discontinued as quickly as possible:

$$\min ITSE = \min \sum_{i=0}^{n} e^2(i) \tag{4}$$

where:

e(i) – discrete series of values representing an error in the control signal.

The parameters, obtained in simulation tests, were introduced to the PLC controlling an operation of one of the machines. The parameters of the controller, used in the second jig, were determined by bringing the collection systems to the stability limit, to the moment when the constant oscillations begin to appear. This was achieved by increasing the K_p gain factor (in a closed system with a regulator) gradually. For such oscillations the critical vibration period T_{kr} was determined. The obtained coefficients were entered into the formulae from Table 2. In both cases the algorithm uses anti-windup system. It prevented an excessive signal increase in the integrator module of the regulator after exceeding the upper or lower limit of the control signal. When these limits are exceeded, an integration stops and takes the last value that occurred before the control signal exceeded a predetermined level. The integrator only starts to sum up the successive error values when the signal returns to the set range [9]. Such a mechanism allows to limit over-regulations resulting from the saturation of the regulator output signal.

Table 2. Optimal controller settings according to the Zigler-Nichols method.

Regulator	Optimal controller settings	
	K_p	T_i
PI	$0.45K_{kr}$	$0.85T_{kr}$

Table 3 compares the relative values of the assumed regulation quality standards in both cases.

Table 3. Values of relative quality metrics for the considered cases.

Regulation quality ratio	$\lvert\bar{e}\rvert$	ITSE
Real object tuning	100.0%	100.0%
Tuning based on computer simulation	106.1%	102.8%

5. Summary and conclusions

Due to an efficient cooperation with the BUDRYK Mine employees, the start-up of new machines was carried out successfully. Introduced changes allowed to improve the functionality and adjust the control system to the user's needs. KOMAG researchers obtained important information on the specific requirements of the user concerning the mentioned system, as well as important measurement data, thanks to which the system will be further improved. It is planned to expand the system introducing additional control functions in the aspect of complex automation of the jig node operation. This applies, in particular, to the collector air supply system.

Basing on the industrial research carried out on the jig control system, it can be concluded that the KOGASTER SSWO control system enables an accurate separation of the processed material. The analysis of the obtained results showed that total losses of flammable material in the waste product were minimal.

The following conclusions can be drawn from the analysis of the jig heavy product collection system and the tests of the control system:

1. The product collection section of the jig is a non-linear object whose dynamic properties change depending on the direction of variation around the work point.
2. The value of the ITSE quality criterion does not change significantly in the analysed variants of regulator parameters selection, which probably proves the appropriate dynamics of the applied control system model.
3. The mathematical model of the collection system proposed in the paper allows for a certain analysis of the jig float collection system as an object of regulation, however, due to the influence of disturbances such as: change of the feed flow rate, its characteristics and change of particular grain classes shares, they cause that the settings are not sufficient for an effective operation of the system in industrial conditions.
4. A further analysis of the effects of using algorithms with different methods of compensation of unfavourable phenomena accompanying the regulation of non-linear processes is required.
5. In further research a more detailed analysis of the collection system is planned, taking into account such disturbances as: change of the feed flow rate, change of particular grain classes shares and change of the feed characteristics, in terms of the structure of the collection system mathematical model.

IOP Conf. Series: Materials Science and Engineering **679** (2019) 012006 doi:10.1088/1757-899X/679/1/012006

6. Although the feed for the jig was transported by one conveyor and despite the position of the separating flap (Figure 1) - the middle position, it cannot be stated that the material was evenly distributed between the two jigs, which could have an impact on the obtained results.

References

[1] Cierpisz S 2012 Automatyczna regulacja procesu wzbogacania węgla w osadzarkach", Ed. Wydawnictwo Politechniki Śląskiej

[2] Jendrysik S and Kost G 2019 Control of bucket conveyor's output *Advances in Intelligent Systems and Computing* Volume **934**, pp 192-200

[3] Nawrocki J 1972 Budowa i eksploatacja osadzarek. Skrypty Uczelniane Politechniki Śląskiej nr 408

[4] Dietrych J 1953 Osadzarki, Ed. PWT

[5] Jasiulek D, Jendrysik S, Kowol D, Łagódka M, Rogala-Rojek J and Woszczyński M 2011 Automatyczny system sterowania i wizualizacji pracy osadzarki pulsacyjnej *Innowacyjne i przyjazne dla środowiska techniki i technologie przeróbki surowców mineralnych. Bezpieczeństwo - Jakość - Efektywność, KOMEKO* pp 129-135

[6] Matusiak P and Kowol D 2019 State-of-the-art pulsating jigs of KOMAG type *Materiały na konferencję: IMTech, Innovative Mining Technologies, Scientific and Technical Conference, Szczyrk, Poland, 25-27 March 2019 s. 1-10, (IOP Conference Series: Materials Science and Engineering 2019 vol. 545),* ISSN 1757-899X

[7] Kowol D., Matusiak P.: Improving the quality of hard coal products using the state of the art KOMAG solutions in a pulsating jig nod. *Materiały na konferencję: MEC 2019, Mineral Engineering Conference, Kocierz, 16-19 September 2019*

[8] Cierpisz S 2012 Automatyczna regulacja procesu wzbogacania węgla w osadzarkach", *Ed. Wydawnictwo Politechniki Śląskiej*

[9] Hanus R, Kinnaert M and Henrotte J L 1987 Conditioning technique, a general anti-windup and bumpless transfer method. *Automatica,* **23**(6), 729–739

IOP Conf. Series: Materials Science and Engineering **679** (2019) 012007 doi:10.1088/1757-899X/679/1/012007

Automatic analysis of video content in the process of monitoring of industrial facilities

M Jura

KOMAG Institute of Mining Technology, Division of Mechatronic Systems, Pszczyńska 37, 44-101 Gliwice, Poland

E-mail: mjura@komag.eu

Abstract. In order to automate the monitoring process, a review of methods of object detection in pictures or video content has been carried out. The methods have been analyzed in terms of their application in the software design, which will be implemented for monitoring with the use of UAVs (unmanned aerial vehicles). The selected methods have been chosen to create prototype software for finding video frames depicting a specific object and archiving the results in relation to this object.

1. Introduction

An adoption time of new technologies in the mining industry is longer than in other branches of industry as the requirements of this industry are strict due to harsh environment of mines. But recently the mining industry increased use of new technologies. Especially a fast adaption can be seen in information technologies as they can often be based on existing hardware. One of the most prominent technologies that will increase efficiency of many industrial processes is artificial intelligence [1, 2, 3, 4]. Another new technology, recently used in mining, is the increasingly popular technique of identifying objects through image analysis. It is used especially in the remote inventory of large industrial installations and stored resources.

Video surveillance is an important part of the monitoring of industrial facilities, processes or machines. It allows to monitor the status of supervised objects without the need to stay close to them. Monitoring systems allow for archiving data and access to it by the user, taking into account the time of registration. Monitoring is conducted continuously or on cycles (e.g. with use of unmanned aerial vehicle), depending on the specificity of monitored objects. A significant disadvantage of video surveillance systems is the large amount of data they generate and the need for analysis by the user. The next step in the monitoring development could be partial or complete automation of the analysis of the collected video content.

In the case of using unmanned aerial vehicles (UAV) in the monitoring process, the obtained video images usually present many monitored industrial facilities, installations or machines. Unambiguous detection and identification of specific devices or phenomena will be possible due to automatic processing of video content by means of a proposed computer program.

Available methods of recognition of objects in images from video files have been reviewed. On the basis of the review and specific requirements for processing video files, the computer application concept has been developed. The individual frames concept also includes an analysis of the evaluating possibility a video file in terms of their use in monitoring.

The prototype software has been developed to detect images showing a specific object and archive the results in relation to this object. The software allows to:

- recognize selected objects,

IOP Conf. Series: Materials Science and Engineering **679** (2019) 012007 doi:10.1088/1757-899X/679/1/012007

- save images (object and time of shooting),
- access saved images by the user.

The prototype software uses programming libraries to process the data received. The OpenCV library is used for processing raster graphics and the ZBar library for decoding barcodes.

2. Review of available methods

2.1. Recognition of keypoints

Keypoint recognition is a computational process designed to determine the abstraction presenting image information and to determine for each point of the image whether that point can be classified as keypoint. Detection of keypoints and comparison of images is one of the important problems in machine vision and robotics. These methods are used in many fields of science and industry. The ideal technique for a detection of keypoints should be resistant to such image transformations as rotation, scaling, brightening, noise and affine transformations. Additionally, the keypoints must differentiate from each other in order to increase the probability of the matches [5, 6].

SIFT (Scale Invariant Feature Transform) is an algorithm for the detection of keypoints created by David Lowe in 2004 [7]. SIFT has proved to be a very efficient for object recognition tasks. However, it requires high computational complexity, which is problematic especially in real time applications [8, 9]. However, there are many modifications to this algorithm that reduce its computational complexity [9, 10, 11]. SIFT allows to solve problems with rotation, brightness, affine transformations and observation point location [7].

SURF (Speed up Robust Feature) is an algorithm very similar to SIFT, but faster and does not decrease the quality of searched points [12]. The DoG (Difference of Gaussians) algorithm used in SIFT has been replaced by a kernel. This is a rough approximation, however, it allows for much faster calculations. SURF uses BLOB detection, which is based on Hessian matrix, to detect keypoints [11].

ORB (Oriented FAST and Rotated BRIEF) is another alternative to SIFT and SURF proposed by Ethan Rublee [13]. ORB is a combination of FAST detector and Binary Robust Independent Elementary Features (BEIEF) algorithm [14]. Initially FAST is used to determine the position of keypoints. Then the Harris Corner Detector is used to limit the number of the points. FAST does not allow calculation of orientation and rotation. It calculates centroids based on the brightness of area with a corner in the middle. The direction from the corner to the centroid gives the orientation. Indicators are calculated to ensure independence from rotation. The BRIEF algorithm does not give good results if there is a rotation in the image plane. In ORB the matrix rotation is calculated according to the orientation of the previously used area and the BRIEF is set according to this orientation.

When considering the detection of markers, first should be considered what kind of markers are to be recognized. It is also possible to approach the problem from the other side and design a marker specifically for a particular task, which could have a positive effect on the performance of that task.

Markers that can be easily recognized with image processing are mainly barcodes. One-dimensional codes have been designed for laser readers in mind, which allowed for the registration of the image in line. Use of matrix barcodes allows to save more data because codes made in two dimensions, increase their bit volume with a square of side length. An additional advantage of some of the matrix barcodes are special tags, which make it easier to find the code and sometimes even its orientation in 3D space. Examples of matrix barcodes are presented in Figure 1. The advantage of using these codes is the availability of libraries enabling decoding them.

IOP Conf. Series: Materials Science and Engineering **679** (2019) 012007 doi:10.1088/1757-899X/679/1/012007

Figure 1. Examples of matrix barcodes.

2.2. Artificial neural networks

Popularity of artificial neural networks has been growing as they can be need for an increasing number of problems to be solved. They most often used for problems that are solved easily by people but do not have satisfactory solutions using classical algorithms. For kind of problems artificial neural networks go through a process of learning with supervision. When the problem is complicated or the human supervision leads to unsatisfactory results, it is recommended to use unsupervised learning. The learning network must be structured in such a way that it can see higher-order abstractions in the input data. Because the networks, then learned in this way do not aspire to a given result, they can detect irregularities and dependencies that were not known before. This makes them useful for analyzing phenomena whose model is incomplete or inaccurate [15, 16].

IOP Conf. Series: Materials Science and Engineering **679** (2019) 012007 doi:10.1088/1757-899X/679/1/012007

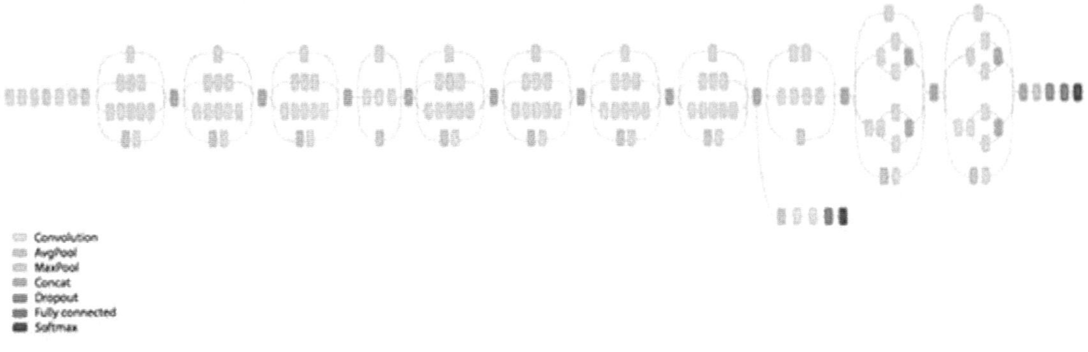

Figure 2. Google Incepcion V3 topology [research.googleblog.com].

Google Inception V3 is a neural network used by Google to recognize objects in images. Figure 2 shows the structure of this network. For the purpose of using it in monitoring, a similar network can be applied but based on a learning set mainly containing monitored objects. A big disadvantage of such a solution would be the need to teach the network every time when adding new objects. However it will be beneficial to use the output of the network, as it will allow to compare several obtained images and to select one that is closest to the one used in the learning set.

2.3. Analysis of a point cloud

Point clouds are sets of points represented in 3D space. Typically each point in a set is assigned three values corresponding to three coordinates X, Y and Z. Depending on the method used, it is also possible to save the color of such a point. Point clouds can be obtained with laser scans or by processing photographs. In case of using UAVs, only the method with the use of photos is available. There is a large selection of software available on the market that calculates point clouds. Examples of such software include Autodesk ReCap (Figure 3), 3DF Zephyr and Selva3D. The software is based on searching for similarities on at least three images and a triangulation to obtain their position in space.

IOP Conf. Series: Materials Science and Engineering **679** (2019) 012007 doi:10.1088/1757-899X/679/1/012007

Figure 3. Autodesk ReCap [autodesk.com].

If point clouds are used, a comparison of scanned objects seems to be the main problem. Some of the programs used to create 3D objects, can also allow to combine these models by finding common features on both models. However, they do not allow to search for smaller objects on the model using this method.

There is also a dedicated software for comparing point clouds. Cloud Compare allows you to compare two point clouds together with their visualization (Figure 4). This software calculates the distances between the two models and presents them as a color model with a legend describing the distances assigned to the colors. Then mean distance and standard deviation are also given. This computer however, does not allow for a fully automatic approach. The approach, used in this application, could also be used in an automatic process.

IOP Conf. Series: Materials Science and Engineering **679** (2019) 012007 doi:10.1088/1757-899X/679/1/012007

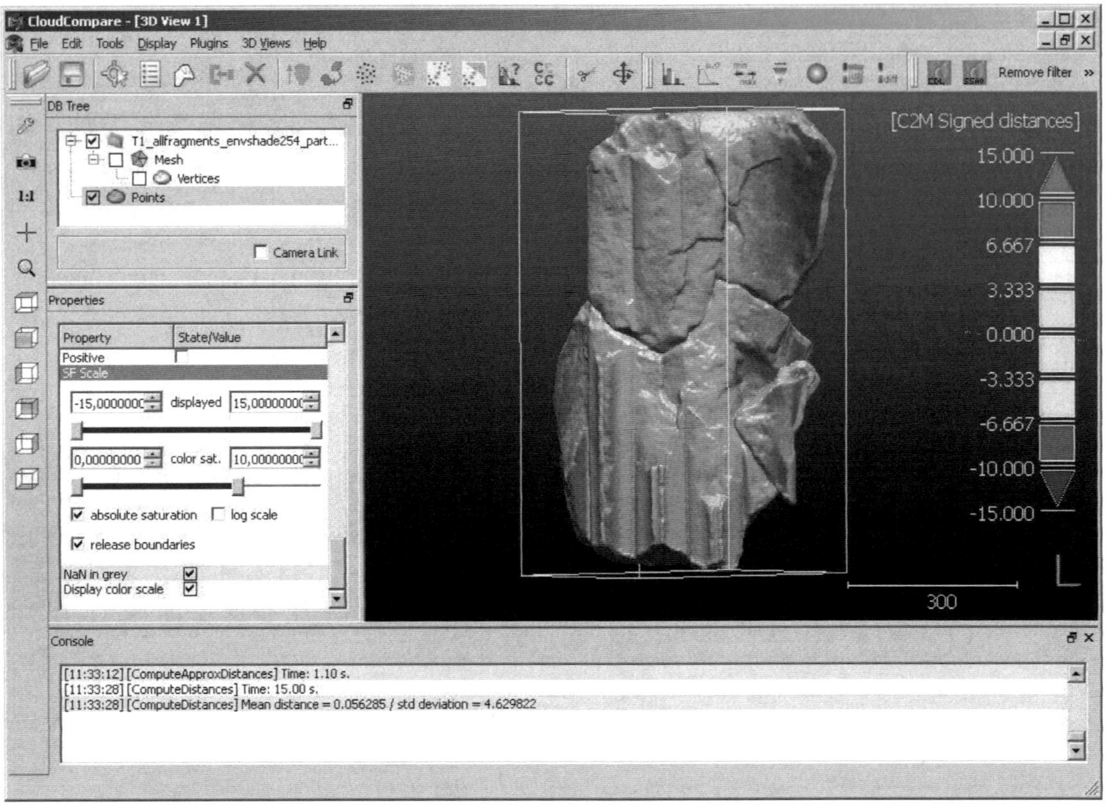

Figure 4. The result of the comparison in Cloud Compare [cloudcompare.org].

It seems possible to use point clouds to search for objects in a scanned model but it requires the creation of your own automated methods for comparing point clouds. An additional limitation is also need for scanning in closed spaces and high computational complexity of the required calculations.

3. Software

3.1. Requirements

A review of available methods of recognizing objects in images serves the purpose of aiding the process of designing software that will be used to monitor the objects selected by the user applying UAVs with video recorders.

The software should allow for an introduction of recorded material in a form of photos or video files. These files can be used entirely by a desktop application or in the case of an application using client-server architecture they can be sent to the server. Subsequent frames of video or photos should be automatically analyzed in search of monitored objects. Due to this, each image will be matched with the information about a detected object and optionally with the information such as perspective, which will allow to choose the best image for monitoring purposes. It is also possible to process output images to match the purposes in a better way. An example of processing can be perspective correction or contrast change.

To speed up development the software should use programming libraries to process the received data. Depending on the selected method of object recognition, it can be OpenCV library for processing raster images or PDAL library for processing point clouds.

When saving images, additional information should also be assigned to the image. The information should contain the name of the object or its subpart that has been chosen for monitoring. It is also necessary to save date and time of registration. This information can be acquired using metadata such as exif or other available methods.

IOP Conf. Series: Materials Science and Engineering **679** (2019) 012007 doi:10.1088/1757-899X/679/1/012007

The computer application user should be able to display saved images by location and time. The images to be viewed should be selected or processed so that they are easier for the operator to compare. This effect can be achieved by selecting the saved images and then processing them.

The user should have a possibility to mark selected areas on the images so that he knows what to pay attention to during the next inventory.

3.2. Solutions

The most important choice in developing a monitoring software is to determine the method of recognizing of monitored objects. It will influence the tools and solutions used.

Image comparison methods using SIFT, SURF, BRIEF, and ORB algorithms are prone to errors related to a rotation of searched objects and their uneven illumination. Used one of these algorithms could involve additional training for the UAV operator to avoid adverse conditions and promote monitoring from a similar perspective for subsequent UAV flights. The software would also require a preparation of images which would allow to identify the monitored objects. Due to the described complications, resulting from the limitations of the analyzed methods, it was decided to use a different approach.

Currently UAVs are often used to create 3D models of physical objects and terrain. This method could also be used to identify and monitor selected objects. However, the available software is a limitation. In the often used available software to create 3D models, it is necessary to fly over the object in a specific way, which can make it difficult to monitor objects in closed spaces or with elements covering the monitored objects. Available methods of comparing point clouds obtained from UAV's overflights, do not ensure an automatic recognition of objects. The use of this method for object recognition would require a development of new solutions and would have a high computational complexity. Use of point cloud comparisons was discarded due to the large amount of work required to obtain working software and predicted cost effectiveness of this solution.

The use of neural networks to detect objects requires a creation of learning sets, which in the case of the detection of objects would be associated with the collection of many images of monitored objects including those that are not subject to monitoring. The learning process of a neural network could be automated, but it would be difficult to predict the effectiveness of such a network. It would be necessary for such a software to be empirically evaluated or for the learning process to be carried out by a specialist. Such a system would require a considerable amount of work with each new application it or some changes in the monitored facilities.

Another possible solution is to use markers that would facilitate a recognition of selected objects. The advantage of such a solution would be the ability of quick adding and removing monitored objects by adding and removing markers. Properly designed markers could also enable a perspective recognition of photographs or videos. If barcodes or other markers are used, only the readability of the marker itself is required, which can be easily assessed by the UAV operator. Due to a high flexibility of the marker-based software, it was decided to use this method in the prototype software.

It was decided to use markers as one of the types of barcodes. It is a proven method of marking objects used in many commercial applications. It was best to use the one that would allow for the most accurate determination of perspective. Some matrix barcodes such as Aztec, QR, Trillcode and Quickmark (Figure 1) have elements designed to facilitate a to determinate of the barcode position. This also allows to determine the perspective and correct it if necessary. Because of the popularity it was decided to use QR codes. This code allows to save the identification number of the monitored object and additional information if necessary.

To read QR codes it was decided to use the ZBar library, which allows to read different types of barcodes. It does not allow to find and correct a perspective. However it is possible to detect and remove the rotation of the code. Due to this limitation decision was made to use OpenCV to detect and correct the perspective. The perspective will also be used to evaluate images from the point of view of their usefulness for monitoring. The position of the code on the image and its scale will also be taken into consideration which will allow to select one image to be saved. Selected images will be placed in a directory named with the identification number of the QR code. The names of image files will contain the time of image registration.

IOP Conf. Series: Materials Science and Engineering **679** (2019) 012007 doi:10.1088/1757-899X/679/1/012007

3.3. Prototype
The prototype is designed to accept a video file recorded during a UAV's overflight as input. During processing input files, it displays a video with additional visualization of QR code detection and data used in the process of saving images. Figure 5 shows one of the moments, when the prototype detected one of markers.

Figure 5. Prototype software during processing video file.

The prototype is able to find markers by detecting contours and nesting information about these contours. Basing on three clearly marked corners of QR code which are shown in purple in Figure 5, the fourth corner at the intersection of the orange lines is calculated. The code detected in this way, is ready for further processing.

The detected code is transformed to remove distortions associated with the perspective. A threshold is used for turning a color image into monochrome graphic to make it easier to read the code. A margin is also added to allow the code to be read by ZBar programming library. The image prepared in this way is shown in the left upper corner of Figure 5.

During an operation, the prototype provides real-time information about detected QR codes. It is displayed in the upper left corner and contains such information as:
- identification number,
- parameter describing yaw and pitch of the code,
- distance of the code from the centre of the image,
- scale of the code compared to the whole image.

IOP Conf. Series: Materials Science and Engineering **679** (2019) 012007 doi:10.1088/1757-899X/679/1/012007

4. Summary

A review of the methods of object detection in images or video content has been carried out. The methods were analyzed in terms of their use in the design of a computer application that was to be used for monitoring with use of UAVs. A method using QR codes to mark monitored objects, was selected. The choice of this method was influenced by its features such as:

- convenience in marking subsequent objects and their removal,
- fast-paced operation of the algorithm,
- possibility of evaluating the angles of yaw and pitch of the monitored object,
- possibility of using the programming library decoding QR codes,
- possibility of ensuring object recognition by ensuring code visibility.

The prototype software has been developed to enable the recognition of previously marked objects in a video file. The prototype allows to evaluate images with detected QR code in terms of their usefulness for monitoring. It allows to reduce the number of recorded data significantly. In the future, the way, in which the images are assessed, may change especially due to the user's opinions. The software introduces the selected images to directories with the identification number on the QR code. File names on the other hand file names contain the timestamp of image capture. The prototype can be used in the future as a part of monitoring system.

The software should be most useful in monitoring of larger objects in large open spaces. Basing on the software capabilities it could be used in big yards storing machinery that are often found around mines or in long industrial installations like powerlines and pipelines.

It is planned to create a user's interface that will facilitate and speed up the work with the software. There will also be a part designed to allow for a legible browsing of saved images with a possibility of selecting elements of monitored objects and adding comments. There are plans to add additional information to the QR code which could improve the software. An example may be an inclusion of the object name in the code so there is no need to enter it in the software manually.

References

[1] Stankiewicz K 3/2016 *A method for the self-organization of a sensor network in belt conveyor exploitation. Maintenance Problems* pp 145-154

[2] Prostański D and Vargová, M 2018 *Installation optimization of air-and-water sprinklers at belt conveyor transfer points in the aspect of ventilation air dust reduction efficiency.* Acta Montanistica Slovaca, 23 (4).

[3] Jasiulek D, Stankiewicz K and Świder J 2011 *An adaptive control system of roadheader with intelligent modelling of mechanical features of mined rock* Journal of KONES Vol. **18** No. 2 pp 197-203

[4] Stankiewicz K, Jasiulek D, Rogala-Rojek J and Bartoszek S 2013 Selected, state-of-the art mechatronic systems in polish underground mining industry ISARC 2013 - 30th International Symposium on Automation and Robotics in Construction and Mining, Held in Conjunction with the 23rd World Mining Congress pp 119-129

[5] Moghaddam B, Nastar C and Pentland A 2001 *A Bayesian similarity measure for deformable mage matching* (Image and Vision Computing) vol **19** nr 5 pp 235-244

[6] Shan B A Novel Image Correlation Matching Approach, JMM, vol 5, nr 3, 2010.

[7] Lowe D G 2004 *Distinctive Image Features from Scale-Invariant Keypoints* (International Journal of Computer Vision) vol **50** nr 2 pp 91-110

[8] Karami E, Shehata M and Smith A 2015 *Image Identification Using SIFT Algorithm: Performance Analysis Against Different Image Deformations* (Canada: St. john's).

[9] Güzel M 2015 *A Hybrid Feature Extractor using Fast Hessian Detector and SIFT* (Technologies) vol **3** nr 2 pp 103-110

[10] Chiu L, Chang T, Chen J and Chang N 2013 *Fast SIFT Design for Real-Time Visual Feature Extraction* (IEEE Trans. on Image Process) vol **22** nr 8 pp 3158-3167

[11] Ke Y and Sukthankar R 2004 *PCA-SIFT: A more distinctive representation for local image descriptors* (CVPR) vol **2** pp 506–513

[12] Bay H, Tuytelaars T and Van Gool L 2008 *Speeded-up robust features (SURF)* (Computer vision and image understanding) vol **110** nr 3 pp 346-359

[13] Rublee E, Rabaud V, Konolige K and Bradski G 2011 *ORB: and efficient alternative to SIFT*

IOP Conf. Series: Materials Science and Engineering **679** (2019) 012007 doi:10.1088/1757-899X/679/1/012007

or SURF (IEEE International Conference on Computer Vision)
[14] Calonder M, Lepetit V, Strecha C and Fua P 2010 *BRIEF: Binary robust independent lementary features* (European Conference on Computer Vision)
[15] Dobrosielski W 2010 *Zastosowanie sztucznych sieci neuronowych do rozpoznawania obrazów* (Studia i Materiały Informatyki Stosowanej), vol **2** nr 3 pp 45-56
[16] Tadeusiewicz R 1993 *Sieci neuronowe* (Warszawa: Akademicka Oficyna Wydawnicza RM) pp 13-14

IOP Conf. Series: Materials Science and Engineering **679** (2019) 012008 doi:10.1088/1757-899X/679/1/012008

Assessment of durability of load-carrying and leveling ropes in selected shafts of underground mines

M Stawowiak[1]*, M Żołnierz[1] and Z Rożenek[2]

[1]Silesian University of Technology, Faculty of Mining, Safety Engineering and Industrial Automation, Akademicka 2, 44-100 Gliwice, Poland
[2]KAZ Serwis Sp. z o.o. Gliwice, Poland

*E-mail: Michal.Stawowiak@polsl.pl

Abstract: The article presents the problem of durability of load-bearing and compensating ropes operated in selected shafts of underground mines. The basic causes of wear of hoisting ropes are described, methods for assessing the wear of wire ropes are given, and criteria for assessing the durability of hoisting ropes are presented. The authors also presented selected results from elongations of lift ropes. The article ends with conclusions resulting from the conducted research.

1. Introduction

Mining hoists are the most important link and are the basic means of transport in the Polish hard coal mining industry. With the help of the hoisting device, the transported extracted material from the underground mines is transported. Lifting equipment also performs a number of other functions, such as the departure and departure of the crew and the abandonment of various materials necessary for the exploitation of the deposit, as well as maintenance of mine infrastructure [4].

The decreasing number of coal deposits lying in favorable mining and geological conditions as well as the continuously increasing depth of exploitation affect the increase of mining costs. If we add to this low coal prices in global markets, the actual picture of hard coal mining in Poland emerges. An opportunity to improve the profitability of exploitation, and thus to maintain the profitability of Polish mines, is an increase in the concentration of extraction. This situation contributes to the need to use increasingly efficient hoisting equipment. This can be achieved by using extraction vessels of increasing capacity as well as speed of movement in the shafts. This entails an increase in the parameters of rope work, which has the effect of reducing their durability [2, 4].

Ropes used in shaft hoists undergo various wear processes. They affect the change of their mechanical parameters as a result of which they lose their operational properties. The ropes are non-linear in the time function of wear processes. The effects of unpredictable wear are attempted to be counteracted by carrying out regular tests and inspections by various methods, from visual to non-destructive ending [4, 8].

To better understand the wear processes of ropes, it may be useful to conduct an analysis of their durability. Ropes working in shaft hoists (as well as materials) of the selected underground mine were subjected to this analysis. The source of data were reports from tests periodically carried out by appraisers, books of periodic inspection of shaft hoists, book of shaft reports.

It should also be noted that due to the large number of test results, including tables and diagrams, the analysis of balance ropes was partially omitted in this article.

2. Causes of rope wear in mining shaft hoists

The main sources of wear of hoisting ropes include corrosion and fatigue cracking of wires. These reasons cause a reduction in the load-bearing cross-section, and hence the weakening of the rope. The phenomenon that intensifies these processes is the clash of external wires. The mechanism of this process is diverse and complex, as a result of which the intensity of this phenomenon varies depending on the type of shaft hoist [7, 8].

In the case of steel - rubber ropes, the main sources of wear include corrosion, which occurs as a result of damage to the rubber jacket. This damage is in most cases caused by falling objects falling into the shaft. Another reason for the damage of the balance ropes are rubbing the rope with elements of shaft reinforcement [8].

- Fatigue wear of steel wire ropes.

Figure 1 shows the case of loading steel wire ropes while working on the wheel drive and during multilayer rope winding on the drum.

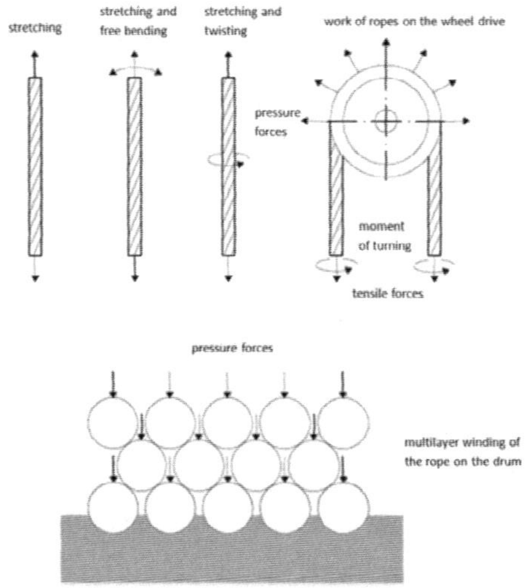

Figure 1. Examples of load on wire ropes [8].

In the case of lifting ropes, the tensile forces are longitudinal forces that change with the lifting height or drawing depth. They appear during braking and start-up, and come from motion resistance, static loads and dynamic forces. In the case of lifting ropes of mining hoisting equipment, the phenomenon of turning the cross-section occurs. It is created as a result of the torque generated from the moment of unraveling in the strands of the rope. Many load phenomena come from bending ropes on propeller wheels. Increased bending stresses occur in the area of running up and running off of wheels and risers [6].

The impact on fatigue consumption for multi-layered winding of ropes on the drums has a high pressure load. During co-operation of ropes with running tracks of rope wheels, excessive pressure causes characteristic fatigue wear. In the first stage, plastic deformations of wires in the form of "lips" are observed. In the next stage, the wires start to crack avalanche. One wire can break in several places and its ends are not separated.

The complex state of stress is of key importance to the pace of fatigue consumption processes. The influence of stretching on rope fatigue consumption depends on the variability and average value of loads, variability and number of cycles, degree of corrosion and rope construction. The influence of bending on fatigue wear depends on the size of longitudinal loads, length of bending, changes in the amplitude of loads, direction of bending and angle of wrap [6].

Wire breakage is a sign of fatigue wear of wire ropes. This phenomenon is possible for visual observation, however, to a limited extent (internal scrap wires).

- Corrosion of lifting ropes.

The form of physical and chemical wear of the carbon steel from which the wires are made is called corrosion. It is the result of the formation of iron oxides, hydrated salt compounds, as well as salts of sulfuric acid. Atmosphere affects the growth of its dullness with a high content of NO_X and SO_X acid radicals, as well as environmental humidity. Corrosion can only occur in the presence of molecular oxygen. The hydration of iron oxides increases their volume as well as the reduction of bending and stretching strength. There are notches on the surface of the wires, surrounded by accumulation of stresses [8].

- Clashes of external wires.

It was found that there are two factors as a result of the action that occurs for clashing of external wires [7]:

- abrasive wear as a result of mutual movement and contact with the occurrence of friction, a pair of elements cooperating with each other,
- fretting consumption.

The abrasive wear may occur as a result of friction between the wheel and the rope when the rope is running up or running down from the thrust or from the rope pulley. There is then slippage of the rope surface relative to the wheel surface [7].

Fretting is called a set of phenomena occurring on the contact surface. This is a multidimensional phenomenon, which is manifested by wear on short parts of the rope. Movements and deformations as well as high stresses are necessary. For freeting to occur, relative slides between interacting elements are necessary. It is often confused with corrosion. A typical sign of freeting is the appearance of short wire thinning [7, 8].

Variable axial force and vibrations cause uneven pressure between the rope and the propeller. At the moment when individual shear forces exceed the friction forces, they slip (Figure 2). As a result of varying pressure between the rope and the groove, both abrasive and free-standing wear may occur [7].

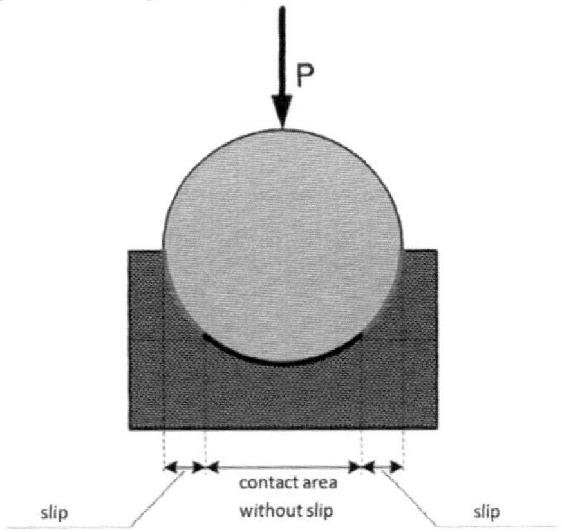

Figure 2. Mechanism of freckle damage [7].

W In the case of shaft hoists with a machine placed on the tower, the rope has contact with only one wheel. As a result, there is no problem with wire clashes. The linings on the wheels have carefully selected material constants, which practically minimizes the clash of wires.

In the case of shaft hoists with the machine located on the framework, clashes occur mainly on the wheels. These wheels are usually without linings. However, the size of the clashes is small and it is not the dominant type of wear.

Shaft lifts with drum machines are characterized by high abrasion of external wires. At the moment of winding the ropes on the drum there is friction between the strands (Figure 3). In the case of multilayer winding on the drum, there is a strong abrasion when the layer is changed by the rope. Abrasion of the rope strands in the groove of the rim causes a rope angle of the rope and the width of the drum [7, 8].

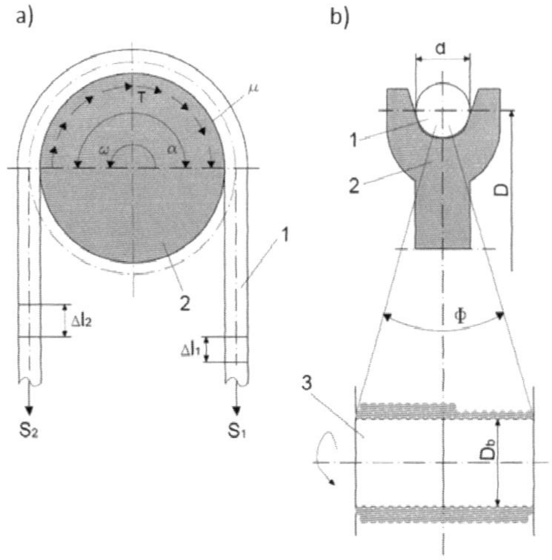

Figure 3. Causes and places of abrasion of steel ropes, where [7]: a) drive wheel, b) drum and rope wheel: 1 - supporting rope, 2 - drive wheel, 3 - drum, D [mm] - diameter of the drive, d [mm] - rope diameter, S1 and S2 [N] - tension in rope branches, T [N] - friction forces, Dl1 and Dl2, [mm] - elongation of rope branch, El - modulus elasticity, α [rad] - angle of wrap on the wheel, m - coefficient of friction, φ [degrees] - angle of the rope on the steering wheels, ω [1 / min] - speed of the wheel.

As a result of abrasive wear, mass losses occur. For drum hoisting equipment there is an asymmetric abrasion of ropes, as shown in Figure 4. A similar phenomenon applies to non-rotating ropes in frictional hoist extracts. Abrasive wear of ship ropes operated in extracts with friction couplings (log and tower) occurs symmetrically.

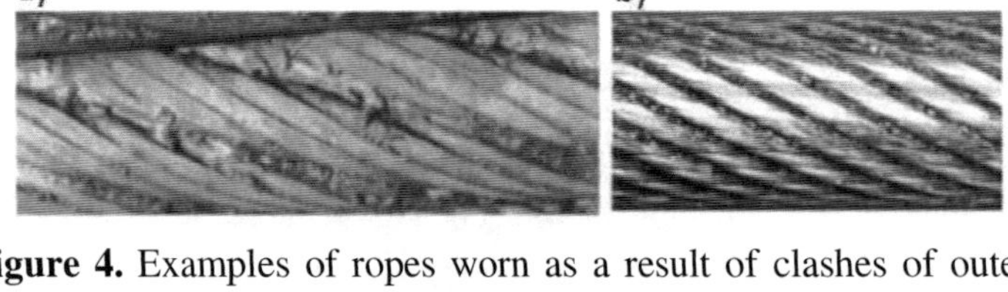

Figure 4. Examples of ropes worn as a result of clashes of outer wires [7]: a) surface of the triangle-nasovian rope with worn outer wires, b) surface of the rope with strands of strands

- Wear due to the action of the torque.

As a result of the occurrence of the unscrewing moment, the ropes wear, which may result in twisting of a part of the cross-section or the whole rope. These phenomena do not occur very often, but almost always have a catastrophic character. They take place in the inner part of the rope, which makes visual observation impossible. They are very dangerous, and the only chance to detect them are periodically repeated magnetic tests [8].

3. Methods for assessing the condition of steel wire wear

In order to evaluate the condition of the rope in operation, several methods are used. Two groups can be distinguished among them [8]:

a) not apparatus - they rely on observing wear processes or on measurements of geometrical features of ropes whose condition changes along with a change in the condition of the rope,
b) apparatus - they rely on the use of specialized equipment. Magnetic tests are the most commonly used.

- Visual method.

This method is based on visual observation of the condition, the available part of the rope. This is the most common diagnostic method for assessing the condition of ropes. It requires a lot of experience of the person conducting the observation. Due to the lack of unambiguous criteria, this method is not very objective. By performing a cyclic counting of fatigue scrap wires, this method can be used to characterize fatigue wear of ropes [8].

Visual observations allow to reveal damage, such as [3]:

- rupture of strands,
- corrosion,
- wire breaks,
- cold rope diameter,
- deformation,
- waviness,
- rope attachment points.

- Instrumental methods - magnetic testing.

Magnetic testing is the most common, apparatus-based method for testing hoisting ropes. The first country that introduced in its mines the legal obligation to test the magnetic lifting ropes of lifting devices was Poland. Today's regulations require magnetic testing of both balance ropes and guide [1, 4, 5].

During the test the rope is magnetized with a permanent magnetic field. The magnetic flux flowing through the rope is generated by permanent magnets. On the fragment of the rope, which undergoes magnetization, an inductive sensor (measuring coil) is located between the pole pieces. The instantaneous value of the magnetic flux associated with the measuring coil changes when the line section on which the damage is located is moved. As a result of this phenomenon, an electromotive force is induced, the value of which is directly proportional to the change of the ferromagnetic cross-section of the tested rope. The influence of this force is also influenced by many factors depending on the measuring head and rope damage parameters. Part of the magnetic flux induced by permanent magnets flows through the space surrounding the rope. This part of the stream is called the stream of distraction. The lines of the scattering flux are parallel to one another, provided that there is no change in the ferromagnetic cross-section of the tested rope on the magnetized section (Figure 5) [6].

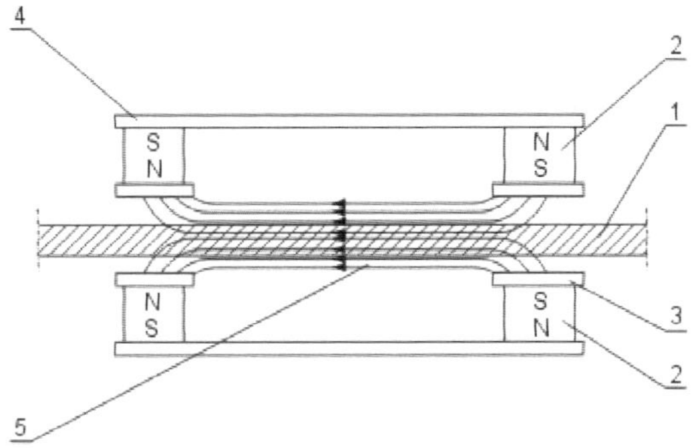

Figure 5. Distribution of the magnetic field lines in the undamaged line, where [3]:
1 - rope, 2 - permanent magnets, 3 - polepieces, 4 - jumper, 5 - stream of dispersion.

IOP Conf. Series: Materials Science and Engineering **679** (2019) 012008 doi:10.1088/1757-899X/679/1/012008

If there is a step change in the rope cross-section (corrosion pits, cracks, etc.) or there is a change in the uniform structure of the rope, the force lines of the magnetic field will be deformed. The value of the magnetizing flux does not change.

For magnetic testing of hoisting ropes, the same method is used - magnetizing the rope with permanent magnets. However, different manufacturers use different types of sensors. Depending on the number and type of detector elements, the design of magnetic concentrators, magnetization susceptibility, sensor operation principles, they provide different signals. Inductive sensors and hallotrons are the most commonly used detection elements. The basic task of the inductive sensor is detection and measurement of step defects. If it is used, the damage should move relative to the sensor. Hall-effect sensors are also used, which give the possibility of measuring the so-called continuous damage such as wire clashes and corrosion. They are also used to increase detection of damages such as corrosion pits, wire breaks [6].

They can be used as external or internal sensors. Data on the components derived from the radial magnetic field force lines are collected by an external sensor. It should be located at a suitable distance from the tested rope immediately next to the armature. Using information from external and internal sensors, it is possible to obtain data on the depth of occurrence of the defect in the tested line. If the defect is deep in the line, the values from both sensors have a similar value. In case when the value of the signal from the external sensor is lower than the value of the signal coming from the internal sensor, the defect is located on the outer layer of the rope. These solutions are considered to be the best for detecting a significant part of typical rope damage [6].

Apparatus for performing magnetic defectoscopy consists of two elements: a measuring head and an output signal recorder. The result of the test is a defectogram on which the test results are registered. The apparatus is equipped with an additional system whose task is to balance the speed changes. The stroke of the tape on which the results are recorded is adapted to the speed of movement of the rope. As a result, the signal coming from the sensors and recorded on the tape does not depend on this speed. The measuring head should be calibrated by an independent person, according to standards, every 3 years.

4. Criteria for assessing durability of load-bearing ropes

The selected criterion for assessing rope durability is the rope time indicator "A" expressed in [MNm/kg]. Ropes with a similar structure, but working in different conditions have a different lifespan. This coefficient is used to compare their "work intensity". For exhaust devices, it is referred to as a formula R. Meebold [1]:

$$A=\frac{N\times Q}{100\times q_i\times i_n}\ [\frac{MNm}{kg}] \tag{1}$$

where:

N – number of work cycles of the statement,

q_i – the weight of the meter of the current carrying rope [kg/m],

i_n – the number of carrying ropes,

Q – maximum permissible load [MN].

The maximum permissible loads come from: the weight of the load rope section from the axis of the rope wheels to the suspension in the upper position of the vessel, the weight of the suspension of the lifting rope, the weight of the suspension rope suspension, the weight of the rolling guides, the weight of the skip or cage, the useful weight of the spoil, the own weight of the trolleys cage), the weight of the balance rope section, measured from the discharge level at the shaft head to the point where the rope returns in the sump [1].

- Sample calculation of the working time indicator for cables operating in the selected mine shaft - eastern range (Figure 6).

Table 1 presents the number of cycles and the value of the work indicator for subsequent sets of ropes working in the eastern range of the selected mine shaft. The following data was used for the calculations [6]:
- maximum static load on the carrier rope: Q = 501 [kN],
- the weight of the meter of the current carrying rope: q_i = 6,9 [kg/m].

Table 1. Number of cycles and the value of the working time indicator for subsequent sets of ropes working in the eastern zone of shaft III [6].

The number of the next set of ropes	Number of lifts cycles	Rope time indicator „A" [MNm/kg]
1	90848	1649
2	92595	1680
3	76159	1382
4	96820	1757

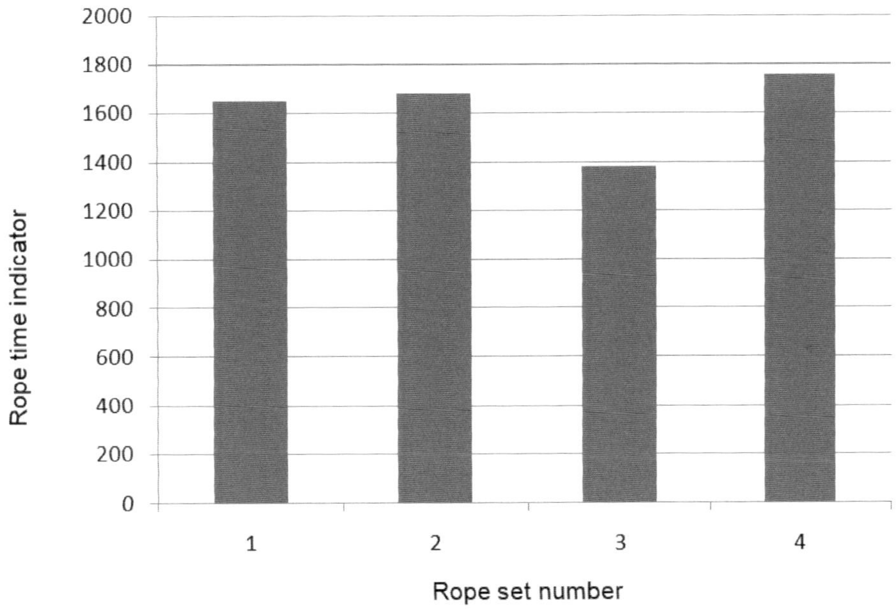

Figure 6. Indicators of working time A for individual sets of ropes, selected production shaft - eastern compartment [6].

- Calculation of the working time indicator for ropes working in the selected mine shaft - west compartment (Figure 7).

Table 2 presents the number of cycles and the value of the work index for subsequent sets of ropes operating in the western range of the selected mine shaft. The following data was used for the calculations [6]:
- maximum static load on the carrier rope: Q = 564 [kN],
- the weight of the meter of the current carrying rope: q_i = 6,9 [kg/m].

Table 2. Number of cycles and the value of the working time indicator for subsequent sets of ropes working in the western range of the selected mine shaft [6].

The number of the next set of ropes	Number of lifts cycles	Rope time indicator „A" [MNm/kg]
1	105483	2155
2	95896	1959
4	101819	2080
5	91386	1867
6	165864	3389

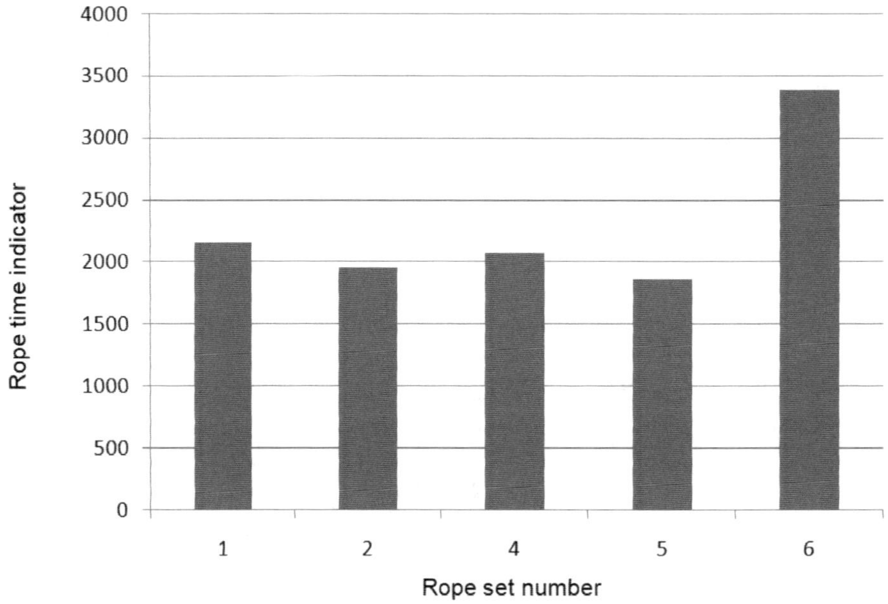

Figure 7. Indicators of working time A for individual sets of ropes, selected mine shaft – compartment western [6].

5. Service life extension of lifting ropes

In order to observe the extension of load-bearing ropes during operation, their elongation is tested as a function of the number of cycles worked. The relative elongation of the rope is expressed by the dependence [8]:

$$\varepsilon = \frac{\Delta L}{l} \times 100\% \quad (2)$$

where:

ε – the relative elongation of the rope [%],
ΔL – elongation of the rope [m],
l – the length of the working rope [m].

The tables below present examples of elongations of selected sets of supporting ropes installed in the eastern and western ranges of the analyzed mine shaft.

- Extending the service life of ropes operating in the shaft hoist of the selected mine shaft - eastern compartment (set 3).

Rope number 1 Ø42,0mm, marked 6x36 WS + FE -Z / z and S / z-n-l-g-1570 during operation was shortened by 3,1 m in total. The rope work as a function of the number of extraction cycles and the working time indicator is presented in Table 3.

Table 3. The course of rope work as a function of the number of lifts cycles, rope 1 set 3 [6].

Rope work time [months]	Number of lifts cycles	Indicator "A" [MNm / kg]	The length of the cut rope ΔL [m]	ε [%]
2	5241	95	1,6	0,18
6	16587	301	1,9	0,21
12	30661	556	2,5	0,28
18	47230	857	2,75	0,30
24	62330	1132	2,95	0,32
27	70457	1271	3,1	0,34

- Extending the service life of ropes operating in the shaft shaft of the shaft III west compartment.

Rope number 1 Ø42,0mm, designated WS6x36 Ap- S / z and Z / s-n-l-g during operation was shortened by a total of 3,3 m. The rope operation as a function of the number of extraction cycles and the operating time indicator is shown in Table 4.

Table 4. The course of rope work as a function of the number of lifts cycles, rope 1 set 4 [6]

Rope work time [months]	Number of lifts cycles	Indicator "A" [MNm/kg]	The length of the cut rope ΔL [m]	ε [%]
1	3902	80	1,8	0,16
6	23718	484	3,00	0,27
12	47440	966	3,05	0,27
18	67654	1378	3,2	0,29
26	94547	1926	3,3	0,30

Figures 8 and 9 show the characteristics of the relative elongation coefficient ε as a function of the number of lifts.

- Shaft III - eastern compartment.

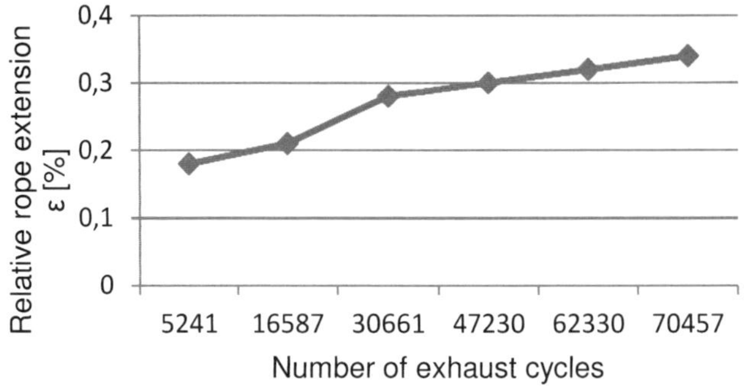

Figure 8. Graph of elongation of the rope ε as a function of the number of exhaust cycles, rope 1 set 3 [6].

- Shaft III - west compartment.

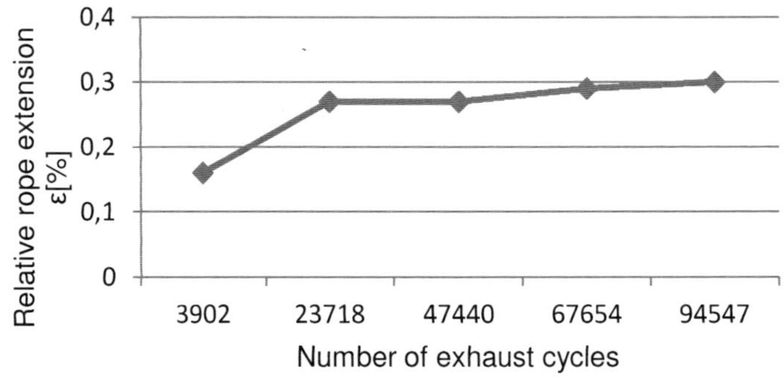

Figure 9. Graph of elongation of the rope ε as a function of the number of exhaust cycles, rope 1 set 3 [6].

6. Conclusions

On the basis of the assessment, the following conclusions can be drawn [6]:

- Support ropes operated in shaft hoists of the analyzed underground mine, they wear mainly due to corrosion. To extend their service life, increase the frequency of relubrication of the ropes.
- Support ropes working in hoisting equipment with the machine located on the frame wear not only as a result of corrosion, but also as a result of wire strokes of the outer layer. This is due to the design of the hoisting device in which the rope runs on the steering wheel at a certain angle.
- Steel-rubber equalizing ropes working in shaft extracts of the analyzed underground mine wear out mainly as a result of corrosion. It comes to it especially as a result of damage to the rubber coating. In order to extend their service life, the period between the detection of damage to the rubber coating and the vulcanization should be shortened.
- Ropes of the same construction, working in the same shaft lift, but greater labor rates are subject to greater elongation.
- In each of the shaft hoists analyzed, the ropes were extended the most in the first 6 months of operation. The elongation is from 62% to 90% of the total elongation.
- Ropes of the same structure, working longer in the same shaft hoist, are more elongated. After working the same number of cycles, the relative elongation is similar.

IOP Conf. Series: Materials Science and Engineering **679** (2019) 012008 doi:10.1088/1757-899X/679/1/012008

References

[1] Hankus J, Szołtysik P et al. *A method of estimating the working time of ropes based on the meganutonometer index.* Central Mining Institute, Katowice 1982
[2] Hankus J *Construction and mechanical properties of steel wire ropes.* Central Mining Institute, Katowice 2000
[3] Hansel J, Kwaśniewski J, Lankosz L, Tytko A *Magnetic testing of steel wire ropes.* Publisher of AGH University of Science and Technology, Kraków 1990.
[4] Kędziora A *Exploitation of shaft hoisting equipment.* Śląsk Publishing House, Katowice 1983
[5] Kwaśniewski J *Magnetic testing of steel wire ropes.* Publisher of AGH University of Science and Technology, Kraków 2010
[6] Mazur S *Analysis of durability of load-bearing and compensating ropes in shaft extracts KWK "Bielszowice".* Diploma thesis, Institute of Mechanization of Mining of the Silesian University of Technology, Gliwice 2015
[7] Tytko A and Nowacki J Abrasive wear of external wires of mining hoist ropes *The International Journal of Transport and Logistics* vol. **10**, 2010, pp. 101-105.
[8] Tytko A *Exploitation of steel ropes.* Śląsk Publishing House, Katowice 2003.

IOP Conf. Series: Materials Science and Engineering **679** (2019) 012009 doi:10.1088/1757-899X/679/1/012009

Tests of BMS Battery Management System with active and passive system of balancing the battery capacity

P Deja

KOMAG Institute of Mining Technology, Division of Drives and Control Systems, Pszczyńska 37, 44-101 Gliwice, Poland

E-mail: pdeja@komag.eu

Abstract. The tests of two BMS Battery management systems, equipped with active and passive systems of balancing the battery capacity, realized within the framework of the HYDKOM 75 project, are discussed in the article. These test were performed on the lithium-ferric-phosphate($LiFePO_4$) battery assembly for the mining mobile machine of BH 3000 B HYDKOM 75 type. The object of tests included two Battery Management Systems (BMS). The first one was a commercially available system most often used in the automobile industry, however the second one was the author's BMS-Battery Management System developed at the KOMAG Institute of Mining Technology within a realization of the HYDKOM 75 project. The tests were conducted both during a battery charging process as well as during discharging the $LiFePO_4$ battery for different values of the load current.

1. Introduction

Tests of two BMS-Battery Management Systems were conducted within the framework of the HYDKOM 75 project [1]. The first system was commercially available (Orion BMS Original) with passive system of balancing the battery capacity, most often used in the automobile industry. However, the second one was the author's BMS management system developed at the KOMAG Institute of Mining Technology with active system of balancing the battery capacity. The tests were conducted on the assembly of lithium-ferric-phosphonate ($LifePO_4$) batteries for the mining mobile machine of BH3000B HYDKOM 75 type. The ideas of BMS systems functioning and their realization of battery balancing processes are discussed in the article. The object of tests is characterized and the tests of two BMS -Battery Management Systems, conducted at the KOMAG Institute, are described.

2. Idea of BMS system functioning

Monitoring and control processes of energy storage in battery assemblies enable their functioning as long as possible, being reliable and stable sources of electric energy. They are characterized by a high efficiency and high level of safety. The diagnostics is realized with use of the BMS-Battery Management Systems. The BMS system, operating correctly prevents any damage to battery, e.g. in the result of its overcharging, excessive discharging or exceeding its temperature limit. The BMS systems also perform a function of a meter of a battery charging level. Besides, they can check a conformity of its characteristics with a receiver requirements and optimize the battery charging process to increase its efficiency. A need of using BMS systems results from an application of new generation batteries, in particular lithium ones, very sensitive to overcharge and excessive discharge.

BMS systems enable to protect assemblies of batteries against inequal charge of its individual batteries as well as against overcharge and excessive discharge. In the case of using a battery assembly composed of several batteries, it is indispensable to equalize their capacities. Equalizing of capacities of batteries consists in equalizing the charge level of all the batteries, which is realized by means of especially designed circuit. Two solutions can be distinguished in this scope: a configuration with a passive or active balance.

2.1. Passive method of battery capacity balancing

An example of the battery capacity passive balancing is shown in Figure 1.

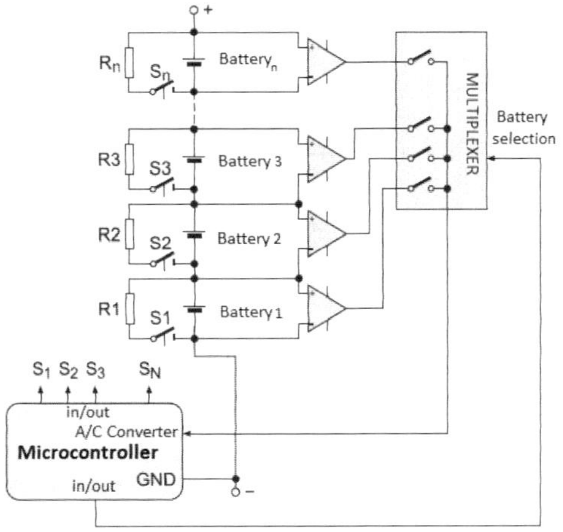

Figure 1. Example of balancing battery assembly capacities [2].

Values of voltages in individual batteries are monitored in the microcontroller by means of the A/C converter, on whose input individual batteries are switched on via a multiplexer. If a voltage of any of the batteries exceeds voltages of the other ones significantly, an appropriate key S is closed. It causes a discharge of the battery by an element of the passive balancing circuit (by a resistor), connected in parallel to each battery and it lasts till the voltage of the overcharged battery is equal to the voltages of the other batteries. Then charging of the packet is continued. Simultaneously, the voltages of the other batteries are controlled all the time.

However, passive balancing of batteries has faults. One of them is a small efficiency resulting from the fact that the energy surplus, stored in the unbalanced batteries, is lost for heat energy in the resistor.

Besides, the total capacity of the battery assembly is limited due to a need of adapting the battery assembly charge level to the capacity of the ‘weakest’ battery.

2.2. Active method of battery capacity balancing

An alternative solution to passive method of battery capacity balancing is the active method. An example of active balancing of the battery assembly capacities is shown in Figure 2.

IOP Conf. Series: Materials Science and Engineering **679** (2019) 012009 doi:10.1088/1757-899X/679/1/012009

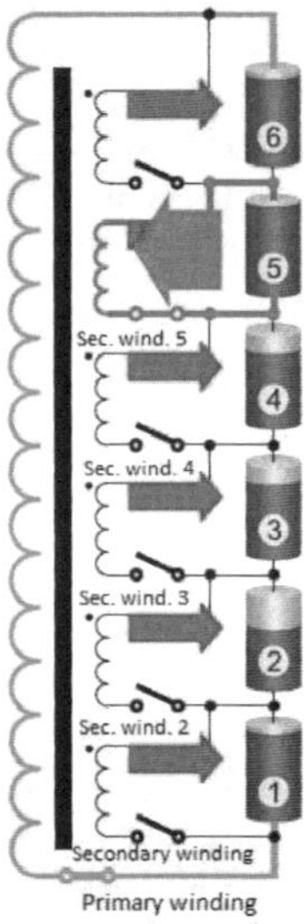

Figure 2. Circuit of active balancing of battery assembly capacities [2].

The active method consists in a conversion of the overcharge from the overcharged battery to the undercharged one (or several). Transistor keys are usually used for this purpose, similarly to the passive method, but instead of resistors an induction coil, being the transformer secondary winding, is connected to the battery.

A reduction of voltage in an overcharged battery consists in its instantaneous connection to the respectful secondary winding of the balancing system, which results in an induction of voltage in the primary winding. The battery key is then opened, but the key of another undercharged battery (or several batteries) is locked. Due to that the energy from the transformer primary winding is transferred to the secondary winding i.e. to the battery.

3. Object of tests

Two BMS systems, were objects of tests. The first one was a commercially available system (Orion BMS Original) with the passive system of battery capacity balancing. However, the second one was the author's BMS – Battery Management System with the active battery capacity balancing system, developed at the KOMAG Institute of Mining Technology.

3.1. Orion BMS Original system with passive system of battery capacity balancing

The Orion BMS Original system with the passive system of battery capacity balancing (Figure 3) is commercially available and commonly used in the automobile industry.

Figure 3. Orion BMS Original with the passive system of battery capacity balancing [4].

The Orion BMS Original is characterized by:

- a possibility of measuring up to 180 batteries connected in series,
- a high electromagnetic noise immunity,
- a possibility of calculating the charge degree (SOC),
- professional interlocking connectors of automobile class,
- a possibility of calculating the discharging current limit (DCL) and the charging current limit (CCL),
- a possibility of measuring voltage of a single battery between 0.5 and 5.0V,
- double interfaces CANBUS 2.0B (fully programmable),
- a possibility of using ODB2 diagnostic protocol.

3.2. BMS – Battery Management System with active battery capacity balancing

The BMS – Battery Management System with the active battery capacity balancing (Figure 4) consists of:

- BMS-S modules, assigned to each battery, are used for measurements of parameters and for a physical realization of the balancing process, i.e. transferring the energy to the other batteries,
- Mounting plates - the plates integrating the BMS-S modules, containing the circuits connected with communication and energy exchange among packets, installed on each packet, do not have any programmable logic,
- BMS-M module – superordinated control system is used for a control of all the constituent modules of the BMS system for a management of the battery operational parameters.

A part of the BMS system, designed for an installation in a single packet, consisting of 16 Lithium ($LiFePO_4$) batteries of 100Ah capacity each.

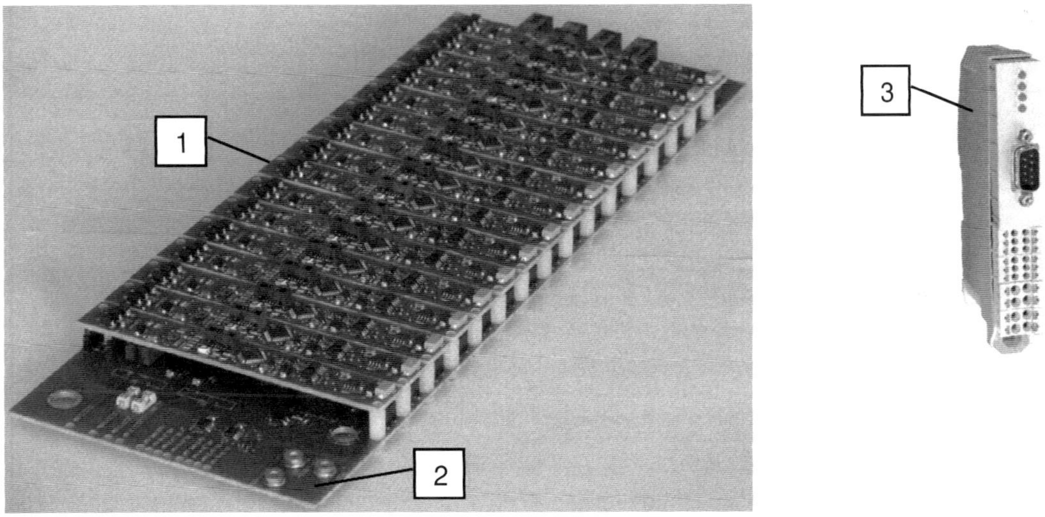

Figure 4. BMS management system with active battery capacity balancing [3] developed at the KOMAG Institute: 1) BMS-S modules, 2) mounting plate, 3) BMS-M module.

The data, among all the modules of the BMS system, are exchanged by means the RS485 interface in series and the MODBUS RTU protocol. The BMS – M superordinated module is equipped with the CAN communication bus.

4. Tests of BMS systems

The tests of the BMS systems, described above, were conducted on the lithium – ferric – phosphate ($LiFePO_4$) battery assembly developed and constructed for the mining mobile machine of BH 3000B

HYDKOM 75 type. The battery consisted of 224 cells connected in series and divided into 14 battery cassettes, 16 cells in each cassette (Figure 5).

Figure 5. Single battery cassette containing 16 lithium – ferric – phosphate cells [3].

Basic technical parameters of the battery, operating together within BMS systems under testing, are presented in Table 1.

Table 1. Basic technical parameters of cell battery.

Number of cells connected in series	224 pcs.
Rated voltage of single cell	3.2V
Maximal voltage of single cell	3.65V
Minimal voltage od single cell	2.65V
Battery rated voltage	716.8V
Battery maximal voltage	817.6V
Battery minimal voltage	593.6V
Rated capacity	100Ah
Charging current	30A
Maximal discharging current	100A

4.1. Tests of Orion BMS Original management system with passive balancing system

The tests of the BMS system with the passive balancing system were conducted during charging and discharging of the battery assembly. The voltages on each battery cell were monitored during the tests. Due to a big number of the battery cells (224 pcs.) two BMS equipment were used in the system. The communication between them was realized by means of the CAN bus.

In Figures 6 and 7 exemplary screen monitoring balancing of the battery cells (for BMS1 – Master and BMS2 – Slave) are presented. The fields marked in red signal balancing of individual cells.

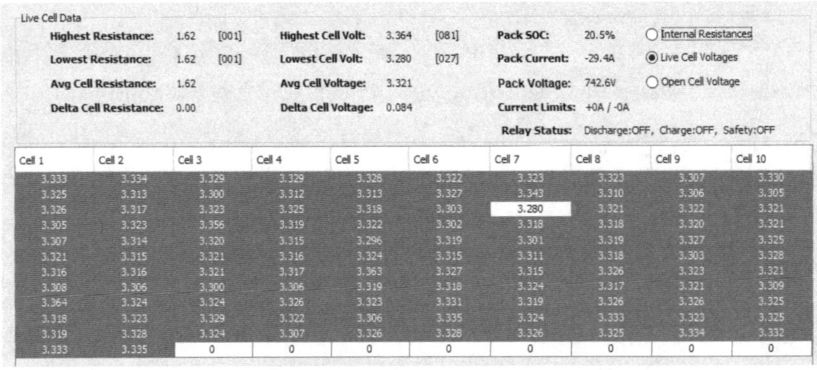

Figure 6. Screen during the monitoring of balancing the BMS1 – Master battery cells [3].

IOP Conf. Series: Materials Science and Engineering **679** (2019) 012009 doi:10.1088/1757-899X/679/1/012009

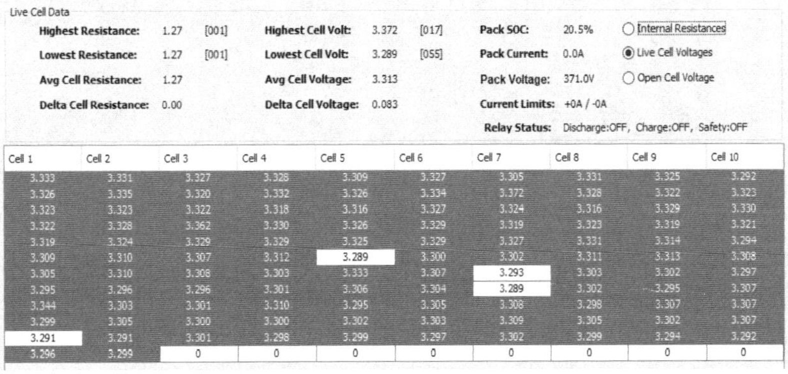

Cell 1	Cell 2	Cell 3	Cell 4	Cell 5	Cell 6	Cell 7	Cell 8	Cell 9	Cell 10
3.333	3.331	3.327	3.328	3.309	3.327	3.305	3.331	3.325	3.292
3.326	3.335	3.320	3.332	3.326	3.334	3.372	3.328	3.322	3.323
3.323	3.323	3.322	3.318	3.316	3.327	3.324	3.316	3.329	3.330
3.322	3.328	3.362	3.330	3.326	3.329	3.319	3.323	3.319	3.321
3.319	3.324	3.329	3.329	3.325	3.329	3.327	3.331	3.314	3.294
3.309	3.310	3.307	3.312	3.289	3.300	3.302	3.311	3.313	3.308
3.305	3.310	3.308	3.303	3.333	3.307	3.293	3.303	3.302	3.297
3.295	3.296	3.296	3.301	3.306	3.304	3.289	3.302	3.295	3.307
3.344	3.303	3.301	3.310	3.295	3.305	3.308	3.298	3.307	3.307
3.299	3.305	3.300	3.300	3.302	3.303	3.309	3.305	3.302	3.307
3.291	3.291	3.301	3.298	3.299	3.297	3.302	3.299	3.294	3.292
3.296	3.299	0	0	0	0	0	0	0	0

Figure 7. Screen during the monitoring of balancing the BMS2 - Slave battery cells [3].

Charging of the battery assembly was conducted at the battery charging rate I_{charge}= 30A DC.

During charging, the voltages on all the cells of the battery assembly were measured. A dozen or so charging cycles were performed. During the tests, it was observed that while charging the first cell, counting from the negative end of the battery cassette, it was different, as regards its parameters, from the other cells in a given cassette, i.e. the voltage was higher than in the case of the other cells in the same cassette.

However, discharging of the battery was performed for different values of the load current intensity ($0.5I_n$, $1.0I_n$, $1.1I_n$). During discharging (similarly to charging), the voltage on the same cell was significantly lower than in the case of the other cells in the same cassette. All the other cells in a given cassette kept similar voltage parameters. Exemplary voltages during charging and discharging of individual cells for one selected cassette of the cell battery are presented in Figures 8 and 9.

Each of the conducted full charging and discharging cycles was performed without any disturbances. No exceeding of equipment permissible operational parameters was recorded.

The BMS management system under testing operated correctly.

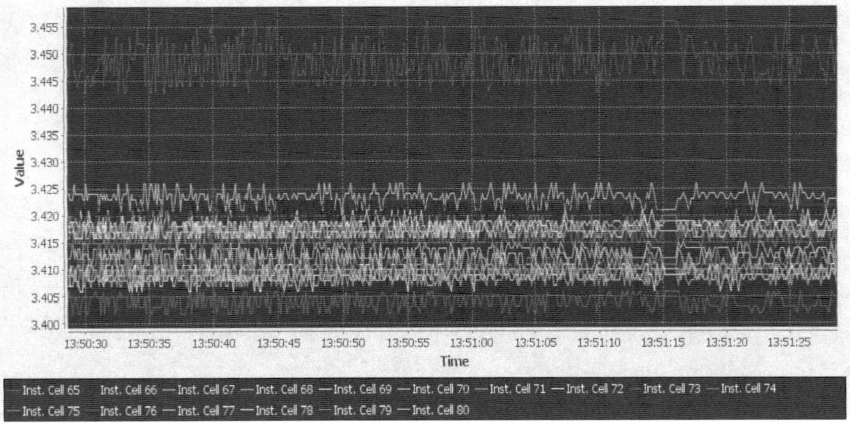

Figure 8. Voltages on the KS5 cassette cells during charging [3].

IOP Conf. Series: Materials Science and Engineering **679** (2019) 012009 doi:10.1088/1757-899X/679/1/012009

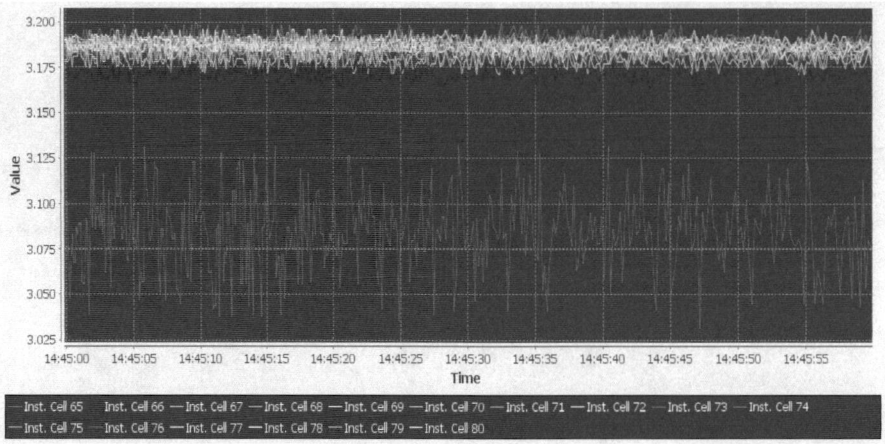

Figure 9. Voltages on the KS6 cassette cells during discharging [3].

In the case of the conducted tests the charging ended when a single cell achieved the maximal voltage, i.e. 3.65V. However, the discharging process ended when a single cell achieved the minimal voltage, i.e. 2.65V. The test results confirmed an efficiency of capacity balancing with the charging current of 200mA and an activation of protections, when the extreme values of voltages and temperatures in the cell battery were exceeded.

4.2. Tests of BMS – Battery Management System with active balancing system

In the next stage some tests of the author's BMS – Battery Management System with the active capacity balancing system were conducted. The tests consisted of two parts. First a calibration of the system at the laboratory stand was performed. Then the BMS system modules were installed on four cell packets (Figure 10). An installation of the BMS system only on the four cassettes was due to the costs of a big number of electronic components indispensable for a construction of the BMS management system.

Figure 10. Four battery cassettes containing 16 $LiFePO_4$ cells with BMS management system and active balancing system of cells [3].

IOP Conf. Series: Materials Science and Engineering **679** (2019) 012009 doi:10.1088/1757-899X/679/1/012009

During the tests an operation of the balancing circuits at full charging and discharging cycles was checked. The tests of the BMS system enabled to adapt the procedure of active balancing in relation to the number and type of cells. The conducted tests showed that modules of the BMS system, installed on the cells, ensure correct balancing of their charges. The tests enabled an empirical determination of the voltage limits on individual cells, at which a realization of balancing (equalization of charges) was most efficient. It can be concluded from the tests that the voltages in individual cells are within the range from 3.23V to 3.30V. The tests confirmed the efficiency of capacity balancing within the charging current 2A which is 10 times higher than in the case of the BMS system with the passive capacity balancing system.

5. Conclusions

The tests of the BMS – Battery Management Systems conducted within the framework of the HYDCOM 75 project are discussed in the article. These tests were realized on the lithium – ferric – phosphate ($LiFePO_4$) cell battery assembly developed and constructed for the mining machine of BH 3000B HYDCOM 75 type. Two cell battery management system (BMS), equipped with the active and passive cell capacity balancing systems, were subject to tests. The commercially available (Orion BMS Original) system, dedicated mainly for use in electric cars, was tested first. The tests confirmed an efficiency of capacity balancing with charging current of 200mA and an activation of the protections when exceeding the limit vales of voltages and temperatures on the battery of cells. The battery managements system (BMS) with the active balancing system, developed at the KOMAG Institute of Mining Technology, was subject to tests as the second one. This system consisted of BMS-S modules installed on the basic plates, enabling a connection to the packet cells and of the BMS-M management module. The management system was developed for four battery cassettes (64 pcs. of single cells). The tests confirmed the efficiency of cells' balancing with charging current of 2A.

From a comparison of both BMS systems it can be concluded that from the point of view of the active cells balancing, where balancing was realized with the current of 10 – times higher intensity, the author's solution is more advantageous. Besides, it should be highlighted that while using the active capacity balancing, the energy stored in the battery is not lost for heat as it occurs in the systems with the passive capacity, balancing system. However, due to an economic aspect of the system construction for the whole cells' battery, it was decided that an implementation of this solution, during the commercialization stage of the project results, will be realized upon a future user's individual request.

References

[1] Kapuściński D and Polnik B Innowacyjny układ zasilająco-sterujący spągoładowarki górniczej, *Maszyny Elektryczne - Zeszyty Problemowe* 2018 nr **2 (118)**

[2] Kurpiel W System nadzoru baterii ogniw akumulatorowych (BMS), *Maszyny Górnicze* 2015 nr **4**

[3] Sprawozdanie z badań. Zadanie 5. Budowa i testy układu zasilająco-sterującego maszyny. Innowacyjna maszyna mobilna z uniwersalnym układem napędu elektrycznego, podwyższającym poziom bezpieczeństwa technicznego. W57.012AY3 (not published)

[4] Website: https://www.orionbms.com

IOP Conf. Series: Materials Science and Engineering **679** (2019) 012010 doi:10.1088/1757-899X/679/1/012010

Test of manoeuvring haulage unit with battery drive conducted within operational trials in the conditions of the mine working

K Kaczmarczyk

KOMAG Institute of Mining Technology, Division of Machinery and Equipment, Pszczyńska 37, 44-101 Gliwice, Poland

E-mail: kkaczmarczyk@komag.eu

Abstract: Suspended manoeuvring haulage units are commonly used for a realization of manoeuvring operations in mine workings. Their application minimizes a necessity of performing manual transportation operations and improves work safety of miners. The PCA-1 battery driven haulage unit, developed at the KOMAG Institute, can be an example of such a haulage unit. Due to lack of experience in using battery-driven equipment of this type in the mine workings, it was indispensable to conduct a series of test, among others, in in-situ conditions. The results of these tests and some observations, made during them, are presented in the article.

1. Introduction

Suspended manoeuvring haulage units are commonly used for conducting manoeuvring operations in mine workings, in material stations or workshops. Their use minimizes a necessity of performing manual transportation operations and improves work conditions of mine personnel. Many solutions of such equipment are available in the market. This equipment, in relation to the drive, can be divided into: electrical, hydraulic, electro-hydraulic, diesel-hydraulic. A novel solution in this group is a battery-driven haulage unit with an electric drive [1, 2].

The PCA-1 suspended battery-driven haulage unit, in which lithium cells $LiFePO4$ are applied, developed at the KOMAG Institute and produced by the Hellfeier Ltd., is an example of such a haulage unit. This equipment, moving on a suspended monorail track, is treated as an assembly of a suspended monorail and it requires an approval of the Chairman of the State Mining Authority [4]

Due to lack of experience in using battery-driven equipment of this type in mine workings, it was indispensable to conduct a series of tests, among others, in-situ tests, in the conditions of the mine working, to confirm an operational safety of the haulage unit.

The tests, mentioned above, were conducted within a trial exploitation in the mine working of the Rydultowy Coal Mine. The test results as well as some observations concerning these tests, are presented in the article.

2. Object of tests – suspended battery – driven haulage unit

A suspended battery-driven haulage unit (Figure 1) was the object of tests. This haulage unit, developed in four versions marked as PCA-1 and PCA-1/ZT, PCA-1/M, PCA-1/ZT/M (PCA-1/ZT and

IOP Conf. Series: Materials Science and Engineering **679** (2019) 012010 doi:10.1088/1757-899X/679/1/012010

PCA-1/ZT/M were subject to tests), is an equipment designed for a transportation of assemblies on a suspended monorail track. The haulage unit can be applied in underground mining plants in the non-methane and methane fields in the workings of "a", "b" or "c" degree of methane explosion hazard and "A" or "B" class of coal dust explosion hazard and in the mining plants of salt and non-ferrous metal ores.

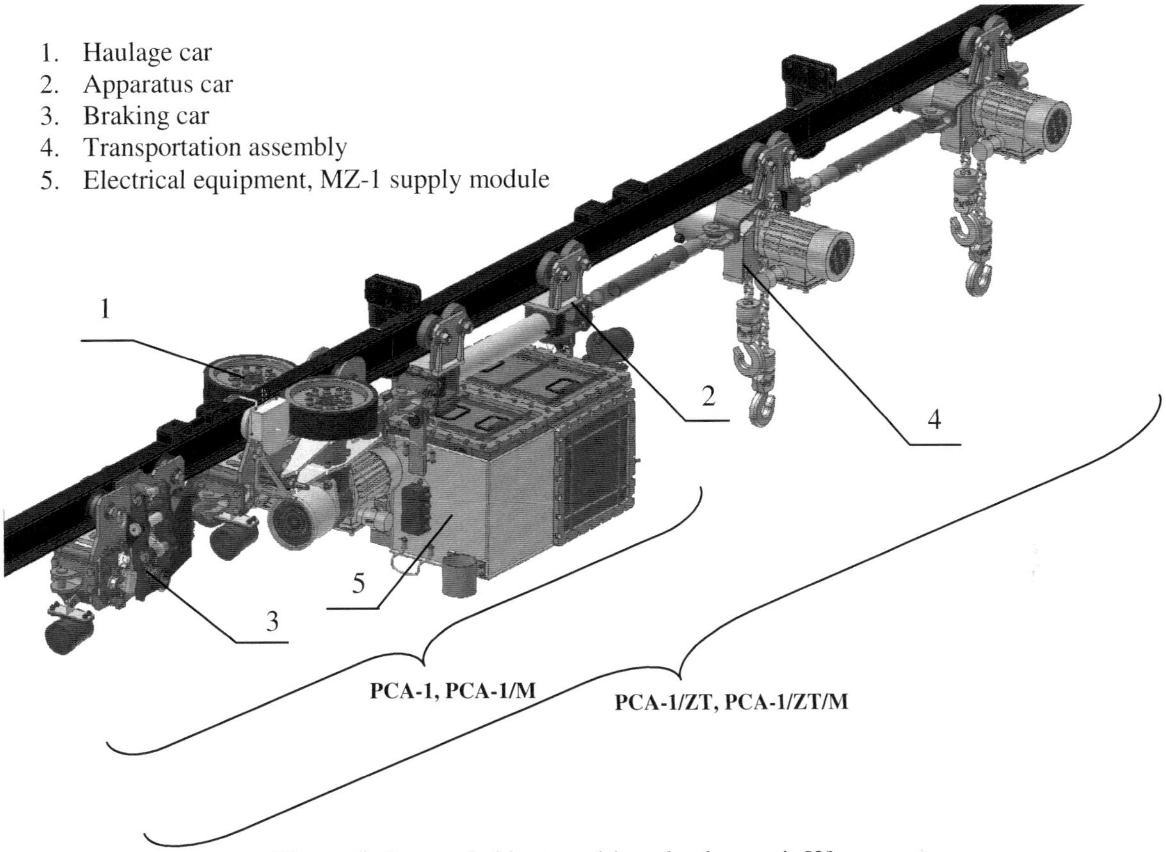

Figure 1. Suspended battery-driven haulage unit [2].

The drive car of the haulage unit is equipped with two frictional gearboxes. A pressure of the wheels to the rail track is realized by means of springs. The drive motors are supplied from a battery of cells installed in the MZ-1 supply module. This module has an apparatus enabling charging of cells and controlling of the equipment operation. In the case of the haulage unit operation on the track inclined above 4°, it is indispensable to use a braking car. A braking car is equipped with a mechanical brake of passive action, i.e. in the case when there is no supply from electric divider, the brake blocks are pressed to the rail web. A brake release is performed in a hydraulic way, using a manually operated pump. An activation of the brake and braking the haulage unit are performed by switching on the emergency switch or automatically in the case of an excessive speed increase, while moving (e. g. at self – rolling down). In the case of haulage units, marked as PCA-1 and PCA-1/M, a collaboration of the brake with the transportation assemblies, equipped with manually operated hoists is foreseen, but in the case of the PCA-1/ZT and PCA-1/ZT/M haulage units a collaboration with the transportation assembly with electric hoists is foreseen [2].

Basic technical parameters of the haulage unit are presented in Table 1.

Table 1. Haulage unit technical data [2].

Parameter	PCA-1 and PCA-1/ZT	PCA-1/M and PCA-1/ZT/M
Haulage unit power(track assembly), kW	2.2 (2x1.1 kW)	
Maximal pull force, [kN]	2.7	4.2
Maximal speed, [m/s]	1	0.5
Motor type	Electric self-braking motors	
Motor supply voltage, [V]	3x230	
Realization of pressure force of drive wheels	By a spring, with a manual control of pressure force	
Battery capacity of MZ-1 module, [Ah]	100	
Possibility of charging batteries	Charging of batteries directly from the flameproof transformer assembly	
Maximal suspended weight on one hook of electric hoist [t]	2.1	

The battery driven haulage unit is a solution, in which a zero – emission battery drive, based on lithium LiFePO4 cells, is used. An application of the integrated charger in this solution enables to charge the battery from any spot in the external mine network, using a flameproof connector being a part of the haulage unit equipment.

It does not require an additional external charger. The battery management system (BMS) supervises an operational safety of the battery cells. The haulage unit was designed in such a way that it enables an adaptation of its composition to the user's needs. In the standard version the haulage unit is equipped with the wire control or optionally with the radio control.

3. Method of conducting tests during the trial operation

Tests in the framework of trials were conducted in the roadway working at the Rydultowy Coal Mine. The tests were conducted on horizontal and inclined segments of the suspended monorail track. The programme of these included four stages [5.6].

- Stage I. Test at standstill.
- Stage II. Operational tests on the horizontal segment.
- Stage III. Operational tests on the inclined segment.
- Stage IV. Tests connected with battery charging.

3.1. Stage I. Tests at haulage unit standstill

Within Stage I of the tests, checking of a possibility of conducting maintenance activities, described in the operational manual, were planned in in-situ conditions. All the activities, made by the operator before starting the work as well as the maintenance operations, were checked. Among others the following activities were checked: an operational correctness of control buttons, switches, circuit – breakers and limit switches.

During the tests the following tasks were to be accomplished:

- a verification of the information in the technical and operational documentation in the scope of check – ups, overhauls and possibilities of acting according to the technical and operational documentation,
- a determination of indispensable time for an everyday overhaul in underground conditions,
- a determination of the method for checking protections,

IOP Conf. Series: Materials Science and Engineering **679** (2019) 012010 doi:10.1088/1757-899X/679/1/012010

- a check – up operation of the control, interlocking and information systems on the LED display against the operational manual,
- a check – up of electric cables to be sure that they will not be subject to any danger hazard during the haulage unit operation,
- a check – up of lighting in the motion direction, the result should be the arithmetic average from 5 tests (the level of visibility/lighting intensity in the distance of 30m),
- a check – up of static breaking force,
- a check – up of pull force (statically), the result should be the arithmetic average from 5 tests.

3.2. Stage II. Operational tests on horizontal segment
The tests, conducted on the horizontal segment included checking of operational parameters and of the functionality on the horizontal segment of the suspended monorail track. Within this stage of tests an operation without any load and an operational with a maximal foreseen load were conducted.

During this stage the following check – ups were performed among others:

- a check – up of the correctness of ground – fault protections activation before starting the tests,
- a check – up of charging level of supply batteries,
- a check – up of operation correctness of motor's brake assemblies working as parking and emergency brakes. The tests were conducted for the following cases: a decay of supply voltage, an activation of emergency switches and an activation of limit switches,
- a check – up of operational correctness of the manoeuvring brake (to the minimal speed) using the V button. The test is aimed at a determination of the minimal time indispensable for the machine braking till it is stopped and a determination of the braking distance,
- a check – up of a possibility of the brakes mechanical release,
- a check – up of operational correctness of individual radio control function,
- a check – up of chain hoist assemblies – a chain hoist at static load equal to 1.25 x nominal load,
- a trip along a horizontal segment with load on the hook equal to 1.1 x nominal load.

3.3. Stage III. Operational tests on the incline segment
The tests conducted within Stage III were aimed at checking the functionality of the haulage unit on the inclined segment of the suspended monorail track. During this stage the following tests were conducted among others:

- a trip on the inclination without any load,
- a trip on the inclination with maximal load,
- a check – up of parking brakes during a standstill on inclination,
- a check – up of braking efficiency using the emergency brake (braking car) on inclination,
- a trip with load on the inclined segment.

3.4 Stage IV. Tests connected with battery charging
Within this stage it was planned to perform battery charging using the transformer assembly of the output voltage 3x230V, equipped with the ZW-40 socket. During these tests the following measurements were taken:

- a temperature measurement of cell battery housing during the charging process,
- a temperature measurement of the flameproof housing on the spot of the converter installation.

4. Test results within the trials at standstill
During the trials conducted in Stage I the activities performed by the operator, before switching the supply were recorded. The operator checked a technical condition of: wires, limit switches, lighting, brake blocks and of oil level in the brake car tank. These activities were performed during one passage

around the haulage unit and did not last more than 1 minute. After switching on the supply, the operator additionally checked battery charge level (on the display), lighting condition, an operation of limit switches and of emergency switch. He conducted a test of ground – fault protections, he checked an operation and tightness of the braking car hydraulic installation as well as an operation of control buttons on the control panel of the haulage unit and of the chain hoists. During the observations no possibility of non – performing the activities, specified in the technical and operational documentation, to be done by the operator, was noticed. All the process of everyday overhaul does not require more time than 10 minutes and it is possible to conduct it in underground conditions. During the tests some check – ups of the leakage protection, using the “test” button, installed in the MZ-1 supply module, were conducted. After pressing it, the CZU, BZ1, BZ2 and BZ3 diodes illuminate. The next start – up of the haulage unit required a reset of the control system, using the reset button. A test of the protections efficiency was finished with a positive result. In the further part of the trials the test of lighting was conducted. After having selected the drive direction in the haulage unit, the white lighting was illuminated automatically in the selected direction and the red one – in the opposite direction. The lighting intensity was measured at two distances of 15 and 30 m from the haulage unit. The intensity of the white light was checked and determined as the average result of 5 measurements. The intensity in the distance of 15 m was 7.9 lux, and in the distance of 30m – 1.0 lux. During the tests, when the haulage unit was illuminated at the distance of 30 m, it was possible to see only the outline of a moving silhouette [7,9]

The following tests concerned operational parameters such as: pull force and static force, braking force. First, these tests were conducted for the haulage unit marked as PCA-1/ZT and after its modernization also for the haulage unit marked as PCA-1/ZT/M (modernization and a development of the PCA-1/ZT/M haulage unit resulted from the observations and test result obtained during the operational trials).

4.1. Measurement of static braking force and of pull force of PCA-1/ZT haulage unit

The tests were conducted with use of a dynamometer installed in a catch of the braking car. The force value was measured in the peak. The measurements were taken in two versions:

- dynamically – without an elimination of clearances in the measurement system,
- statically – eliminating clearances in the measurement system.

The pull force in the dynamic version was 3.380 kN. It was the value obtained for the haulage unit marked as PCA-1/ZT. This haulage unit is characterized by the mechanical ratio i=40. The measurement was also taken at the initial tension of the measurement system, aiming at eliminating an impact of dynamic forces. A measured value of the force, in the static version, was 2.7 kN. Static measurements of the pull force were conducted until the protections were activated and the haulage unit was switched off automatically. The measurements of the static braking force were conducted separately for the braking car assembly and for the drive car. A dislocation of these assemblies, during the measurements, was realized by means of a manual chain hoist. During the measurement a tension of the springs was increased to determine the tension enabling to obtain the braking force exceeding 1.5 x pull force. The obtained braking force for the braking car and for the haulage unit exceeded 4.05 kN [8].

4. 2. Measurement of static braking force and of pull force of PCA-1/ZT/M haulage unit

During a realization of the pull force measurement in the ‘dynamic’ version, the haulage unit advanced about 40 mm till the measurement system was tensioned. In this case the pull force varied within the scope from 4.5 to 4.8 kN till the drive system was automatically switched off by the protective system.

During the static measurement, the system was initially tensioned to eliminate an impact of dynamic forces on the measurement results. The valve of the recorded pull force was 4.42 kN. The measurement of the pull force in static conditions reflects a situation of a start – up on an inclination uphill after stoppage. A measurement of the static braking force was realized as for the PCA-1/ZT haulage unit, however, in this case maximal, possible to be obtained braking force, was searched additionally. The obtained braking forces were: for the drive car – 9.05 kN, for the braking car – 6.9 kN. The obtained values enable to state that for the finally accepted pull force, in the case of PCA-1/ZT/M haulage unit, which equals 4.2 kN (Table 1), it is possible to obtain the static braking force which equals 6.3 kN. The method of conducting measurements is shown in Figure 2.

Figure 2. Measurements of pull force and of static braking force.

5. Test results within operational trials on horizontal segment

The tests, realized within this stage, were connected with the haulage unit trip on the horizontal segments of the suspended monorail track. It was observed that a drive with the maximal haulage unit speed of 1 m/s (foreseen for the PCA-1/ZT/M haulage unit), may cause problems to the operator, who moves along the roadway working, to observe the advancing haulage unit and the transported weight and simultaneously the track on which it moves. An adaptable possibility of speed control enables the operator to adjust the trip speed to the conditions on the transport track. During the trip, efficiency tests of the emergency braking system were conducted in the conditions of: the dead – man's handle release, an activation of the emergency switches. An operational correctness of the control function, using the wire and radio control, was tested as well. The result was positive. There were tests aimed at a determination of the braking path length in the case of manoeuvring braking. The manoeuvring braking is realized by pressing and maintaining the pressure of the V button on the operator's panel. In such a case a speed reduction of the drive motors, by the converter controlling their operation, is caused. Such braking does not stop the haulage unit but only reduces its speed to the minimal value. A complete stop happens after releasing the dead-man's button on the control panel. On the base of these measurements, the braking path was determined, starting from the maximal speed developed by the haulage unit, which varies from 5.5 m to 6 m. An exact determination of the path length was not possible due to the fact that the operator assessed the point of reaching the minimal speed by the haulage unit subjectively. During the observations of the haulage unit operations, conducted by the mine personnel, frequent stops of the haulage unit, due to a release of the dead-man's button, which caused their instant stop using the brakes installed on the drive motors, were noticed. Braking, due to a reduction of the rotational speed of the motors during a normal haulage unit operation, requires the operator's appropriate earlier reaction. During operational tests temperature measurements on the surface of the battery housing in the MZ-1 supply module, on the motor fame and on the gear - box frame were conducted. The measurements were taken during a half – hour haulage unit trip along a trial segment. The test result were as follows: battery housing – 32.9°C, gear – box frame – 27,4°C, motor frame – 34.2°C. The hottest surface was the motor frame. In the framework of these trials chain hoist assemblies were subject to testing as well (Figure 3). The tests consisted in hanging the weight of 1.25 x nominal load on the hook and observing if a self – lowering of the weight happens during the tests of raising and lowering the weight equal to 1.1 x permissible load. In both cases no abnormalities were seen [7,8].

Figure 3. Testing of chain hoists.

6. Test results of operational trials conducted along inclined track segment

Trials on the inclination were conducted for the PCA-1/ZT/M haulage unit on two inclined rail track segments of 2 m length, in which the inclination angle of individual rails was not constant and varied from 8° to 12°. The trip tests without any load were conducted first. During a trip downhill possibilities of trip speed change and of a stop, including an emergency stop, were checked. No events, indicating an incorrect operation, were observed. During the tests with load on the hooks of the electric hoists a flameproof contactor circuit – breaker of the weight about 290 kg was hung. The conducted tests aimed at checking the haulage unit operation at a start – up uphill after having stopped. During a standstill on inclination, the haulage unit was maintained by the parking brake installed on the motor, which, through the drive system, generated the braking torque on the drive wheel. While setting the haulage unit in motion the voltage was applied to the motor winding and simultaneously the brake was released. In the case of the excessive loading of the haulage unit, the motor was not capable of generating the torque on the transfer shaft fast enough which might cause that the shaft would rotate in the opposite direction than the applied one. During the tests of the start uphill, in the first stage the haulage unit moved in the opposite direction at the distance of max 10 mm, then it advanced uphill.

7. Test results of operational trials connected with battery charging

The MZ-1 supply module of the haulage unit was equipped with the charger integrated with the converter and the flameproof connector, which enabled to charge the battery in any place, from the flameproof transformer assembly, equipped with the connecting socket compatible with the plug. Due to the fact that the heat was released while charging the battery, the conducted tests consisted in temperature measurements of the MZ-1 external surfaces during the charging process. The temperature of the MZ-1 module external surface was 33.5°C before starting to charge on the spot of the battery installation, and on the spot of the converter it was 33.4°C. During the tests the battery of the MZ-1 module was charged with the voltage of 225V and the current of 2.5A from the OZTK 0632 assembly. After 45 minutes of charging, the temperature of the measured surface stabilized on the level of 39.5°C for the battery housing and 43.2°C for the module surface on the spot of the converter installation. The conducted measurements enabled to identify the MZ-1 module surface which was heated most [7].

IOP Conf. Series: Materials Science and Engineering **679** (2019) 012010 doi:10.1088/1757-899X/679/1/012010

8. Summary

The tests of the suspended battery driven haulage unit were conducted within a trial exploitation required by a decision of the Chairman of the State Mining Authority and it was indispensable for obtaining an approval for an application in mine workings. According to the above mentioned decision the mover, Hellfeier Ltd. together with the KOMAG Institute elaborated a programme of trials, which was approved by the OBAC Certifying Body. The tests, within the trials, were conducted in the Rydultowy Coal Mine workings.

The conducted tests delivered a lot of information concerning a functionality and operational possibilities of the haulage unit. They enabled to verify the operational parameters of the haulage unit as regards its use. Due to that a determination and a verification of , among others, pull force, transportation capabilities, degree of heating external surface of the haulage unit and lighting efficiency in mine conditions were possible. The tests confirmed a possibility of a safe application of battery supply based on the LiFePO4 cells, which were installed in the MZ-1 module. The information, obtained from the mine personnel operating the haulage unit, turned out to be valuable. Among others, basing on this information a design modernization of the haulage unit, consisting in increasing the mechanical ratio in the drive system, was introduced. In the haulage unit, marked as PCA-1/ZT/M, the pull force was increased to 4.2 kN at a simultaneous reduction of the maximal speed to 0.5 m/s and at maintaining the same power as in the PCA-1/ZT haulage unit. The conducted tests also delivered a series of other pieces of information which will enable a further development of the equipment.

References

[1] Budzyński Z, Kaczmarczyk K and Pieczora E Akumulatory nowej generacji w górniczych systemach transportowych na przykładzie ciągnika PCA-1. *Napędy i Sterowanie* 2012 nr **2** s. 108-111

[2] Dokumentacja Dopuszczeniowa. Podwieszony Ciągnik akumulatorowy PCA-1/ZT. Styczeń 2019 (not published)

[3] Konsek R, Kaczmarczyk K, Budzyński Z and Polnik B Podwieszony ciągnik akumulatorowy PCA-1. *Napędy i Sterowanie* 2013 nr **7-8** s. 112-115, ISSN 1507-7764

[4] Obwieszczenie Marszałka Sejmu Rzeczypospolitej Polskiej z dnia 4 kwietnia 2019 r. w sprawie ogłoszenia jednolitego tekstu ustawy - Prawo Górnicze i Geologiczne

[5] Program Prób. Podwieszony ciągnik akumulatorowy PCA-1/ZT. W26.085.02AY wyd.1 (not published)

[6] Program Prób, aneks nr 1. Podwieszony ciągnik akumulatorowy PCA-1/ZT. W26.085.02AY wyd.1. Aneks nr 1 (not published)

[7] Sprawozdanie z badań nr LL/290/2017. Gliwice, 2018 (not published)

[8] Sprawozdanie z nadzoru autorskiego. Podwieszony ciągnik PCA-1, PCZ-1/ZT. W26.08.01NA5, W26.085.02NA5. ITG KOMAG 2018 (not published)

IOP Conf. Series: Materials Science and Engineering **679** (2019) 012011 doi:10.1088/1757-899X/679/1/012011

Testing the possibility of using a selective crushing process for selected types of hard coal

D Kowol[1]* and P Matusiak[1]

[1] KOMAG Institute of Mining Technology, Division of Preparation Systems, Pszczyńska 37, 44-101 Gliwice, Poland

*E-mail: dkowol@komag.eu

Abstract. The article describes the design solutions of KOMAG drum crushers. In the case of significant differences in hardness of coal and gangue, these devices are the most appropriate and the cheapest means of preliminary beneficiation of run-of-mine. The results of laboratory tests of susceptibility to selective crushing of selected hard coal types by the drop method are presented. The possibilities of using a drum crusher for deshaling the analyzed materials are determined.

1. Introduction

Hard coal run-of-mine contains a significant amount of gangue, especially in thick grain grades >80 (100) mm. The quantity of stone reaches a value of 30 to even 70% in the grain class> 50 mm [1, 2].

In order to obtain commercial assortments with desirable quantity and quality parameters of hard coal, it is necessary to undergo a series of appropriate processing processes.

The first of them is preliminary deshaling of run-of-mine, which can be done using the differences of coal and gangue density. It takes place in a pulsating water jig [2, 3, 4, 5, 6] and beneficiated with heavy liquid [2, 7] (so-called "wet" method), or using selective crushing [2, 8, 9, 10] (so-called "dry" method).

The purpose of deshaling is to prepare raw material for further beneficiation processes, while also extracting some of the large stone grains from it.

The idea of the "dry" method, known in the world mechanical processing since the early 1950s, consists in dropping the coal and stone grains from a proper height onto a hard surface. Coal and midlings grains, as less compact, break down during the fall. Concise stone grains remain intact.

The device for the process of selective crushing of the run-of-mine is a Bradford type drum crusher, which simultaneously performs processes of grinding, classification and beneficiation. The Polish equivalent of this device is KOMAG type KB drum crusher.

KOMAG KB series drum crushers are most often used for selective crushing of coking coal, characterized by significant susceptibility to crushing, in coal mechanical processing plants, equipped with water pulsating jig for a beneficiation in the broad grain class of 70 (60) -0.5 mm.

Due to the fact that the technology of stone separation in the process of selective crushing requires a significant difference in the compactness of stone and coal, the decision on its use should be preceded by tests of susceptibility to crushing of the potential feed.

IOP Conf. Series: Materials Science and Engineering **679** (2019) 012011 doi:10.1088/1757-899X/679/1/012011

The assessment of susceptibility to selective crushing should be determined on the basis of specific parameters of the run-of-mine obtained during the tests of the susceptibility of a given material to spontaneous crushing under the influence of gravity. One of the possibilities of estimating it is the "drop" method, which consists in dropping individual grains of coal and stone into a metal plate from the height close to the assumed diameter of the crusher drum.

The purpose of the article is to present the possibilities of using KB crushers in Polish processing plants.

2. Drum crushers for selective crushing of run-of-mine

Bradford type drum crushers (KOMAG) are devices using the force of gravity and the hardness difference of individual grains intended for selective crushing of hard coal. Figure 1 shows the KB 3200x6000 drum crusher.

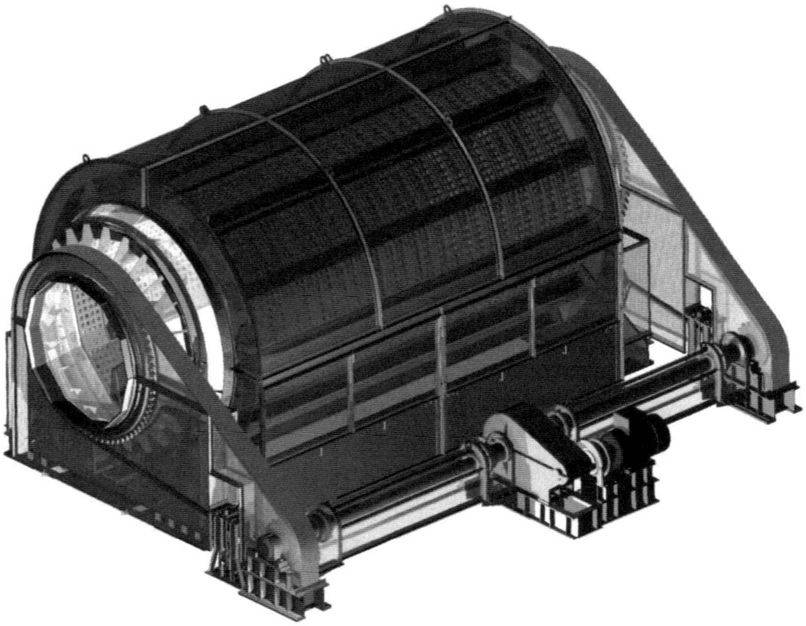

Figure 1. KB 3200x6000 drum crusher.

The main component of the drum crusher is a drum with a horizontal axis of rotation, consisting of a cylindrical part and an inlet and outlet funnel. The cylindrical part of the drum is lined with sieve plates, and the size of the holes in the sieves depends on the technological requirements.

Inside the drum there are shelves, whose task is to move the material along the axis of the drum, while moving it up. Crushed grains, most often coal, leave the device through the holes in the lateral surface, while hard grains, most often stone, are transported along the axis of the drum and leave the device through the outlet opening at its end [10].

Table 1 lists the technical data for a series of drum crushers of KOMAG type KB.

Table 1. Technical data for drum crushers of KB type [11].

Parameter	Unit	KB 2600x4000	KB 3200x5000	KB 3200x6000
Capability	t/h	to 600	to 1000	to 1200
Diameter x length of the drum	mm	2600x4000	3200x5000	3200x6000
Diameter of plate openings	mm		50, 80, 120 or 200	
Drive power	kW	45	55	55
Rotational speed of the drum	min^{-1}	10.2 or 6.7	9.7 or 6.2	9.7 or 6.2
Weight	kg	31500	50500	56000

Crushing machines are affected by a number of factors, among which the following ones can be distinguished:

- the way of preparing the feed,
- diameter and length of the drum,
- number and location of shelves,
- the size of the holes on the surface of the drum,
- rotational speed of the drum.

Although the process, carried out in drum crushers, is complicated and depends on a number of variables, these crushers have a number of advantages such as a very long life, an easy use and maintenance which in effect result in low machine operation costs.

At the KOMAG Institute of Mining Technology, different drum crusher versions have been developed enabling the "dry" deshaling process to be carried out directly in underground mines. The construction solutions facilitate transport and assembly in underground conditions. Assuming more difficult working conditions of the crusher in underground conditions, the drive power was increased to 75 kW [10, 11].

3. Research and tests at KOMAG

3.1. Methodology and scope of research project

Crushing susceptibility tests were performed for two types of coal from three mining plants being parts of the Ruda Mine:

- Bielszowice - type 34.2 (gas and coking coal),
- Halemba - type 34.2 (gas and coking coal),
- Pokój - type 33 (gas coal).

The tests of the run-of-mine samples were carried out at the KOMAG laboratory, using a test rig equipped with a 75 mm thick steel plate, a 810 mm high cover, and an elevation allowing the grains to be crushed by drop [12].

During the tests, using the drop method, the samples of coal and stone were dropped at random from the height of 2 m. Figures 2-4 present examples of coal and stone from the tested types of coal. Before the crushing tests, the weight and size of the primary grains> 80 mm were determined.

IOP Conf. Series: Materials Science and Engineering **679** (2019) 012011 doi:10.1088/1757-899X/679/1/012011

Figure 2. Grains of coal (a) and stone (b) of spoil from Bielszowice Mine.

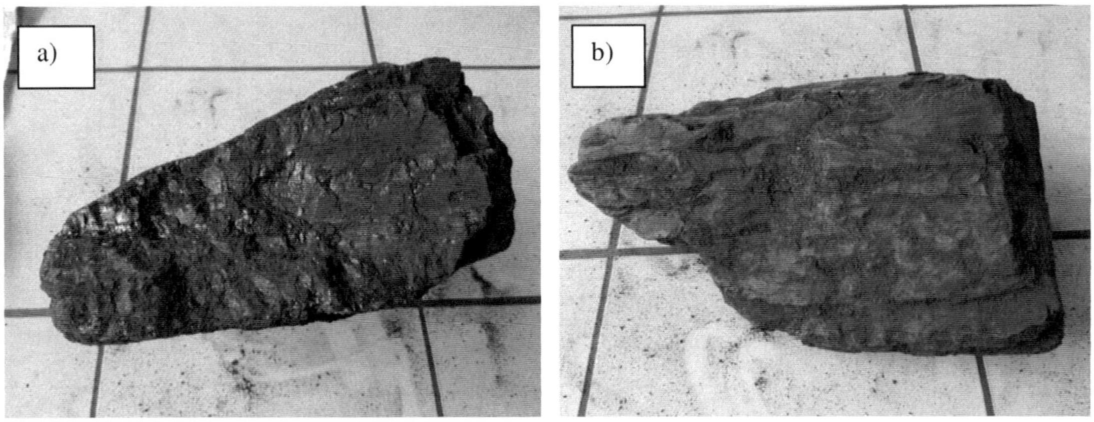

Figure 3. Grains of coal (a) and stone (b) of spoil from Halemba Mine.

Figure 4. Grains of coal (a) and stone (b) of spoil from Pokój Mine.

During the material crushing tests a series of single grains discharges were made. The series means a multiple of one grain drop> 80 mm until it is completely crushed to grains below 80 mm. The maximum number of discharges was 30.

After each drop, grains <80 mm were sieved and the grains above this size were dropped until they crushed below 80 mm or up to 30 discharges. After every 5 drops the weight of these grains was determined.

3.2. Test results

Bielszowice Mine

<u>Material characteristics</u>

The weight of coal grains submitted to crushing ranged from 4 kg to 20.2 kg and their sizes varied from 18 x 17 x 14 cm to 33 x 33 x 24 cm. The weight of stone grains ranged from 11.4 kg to 17.6 kg and their sizes varied from 25 x 24 x15 cm to 45 x 27 x10 cm.

<u>Tests of dropping coal</u>

The smallest number of discharges required to break up the coal grains completely was 7. In seven individual trials coal was completely crushed to <80 mm after 7, 12, 14, 16, 17, 19 and 23 discharges. In two trials, coal was broken up after 15 discharges. In a single sample, the grain was not completely crushed after 30 discharges, and 2.8% of the grains remained. The average number of coal discharges during crushing was 16.8.

The results of coal discharges are summarized in Table 2. The W_z symbol indicates grain dimensions, while M_z - its initial weight.

Table 2. Tests of dropping coal – Bielszowice Mine [12].

Sample 1 W_z=31x31x19 cm M_z=14.39 kg		Sample 2 W_z= 28x22x20 cm M_z= 9.19 kg		Sample 3 W_z= 38x24x22 cm M_z= 15.99 kg		Sample 4 W_z= 33x33x24 cm M_z= 20.18 kg		Sample 5 W_z= 29x27x13 cm M_z= 8.05 kg	
Number of discharges	Weight, %	Number of discharges	Weight, %	Number of discharges	Weight, %	Number of discharges	Weight, %	Number of discharges	Weight, %
0	100.0	0	100.0	0	100.0	0	100.0	0	100.0
5	68.3	5	39.9	5	58.4	5	36.4	5	66.1
10	26.1	10	13.9	10	9.4	10	9.0	10	39.6
15	11.7	12	-	14	-	15	-	15	18.6
20	4.9							19	-
25	3.9								
30	2.8								

Sample 6 W_z= 32x23x14 cm M_z= 9.42 kg		Sample 7 W_z= 32x28x20 cm M_z=15.78 kg		Sample 8 W_z= 18x17x14 cm M_z= 3.96 kg		Sample 9 W_z= 33x28x19 cm M_z= 14.87 kg		Sample 10 W_z= 28x28x17 cm M_z= 7.14 kg	
Number of discharges	Weight, %	Number of discharges	Weight, %	Number of discharges	Weight, %	Number of discharges	Weight, %	Number of discharges	Weight, %
0	100.0	0	100.0	0	100.0	0	100.0	0	100.0
5	26.1	5	27.8	5	11.6	5	35.5	5	35.2
10	10.4	10	6.8	7	-	10	17.5	10	16.9
15	-	15	4.2			15	0.6	15	5.0
		20	3.9			16	-	17	-

<u>Tests of dropping stones</u>

In none of the 5 trials, the stone below 80 mm after 30 discharges was crushed. Weight loss of stone over 80 mm was about 26%. The results are shown in Table 3.

Table 3. Tests of dropping stone – Bielszowice Mine [12].

Sample 1 W_z=24x22x14 cm M_z=17.60 kg		Sample 2 W_z= 25x24x15 cm M_z= 11.37 kg		Sample 3 W_z= 27x24x15 cm M_z= 19.61 kg		Sample 4 W_z= 31x25x11 cm M_z= 14.99 kg		Sample 5 W_z= 45x27x10 cm M_z= 12.71 kg	
Number of discharges	Weight, %	Number of discharges	Weight, %	Number of discharges	Weight, %	Number of discharges	Weight, %	Number of discharges	Weight, %
0	100.0	0	100.0	0	100.0	0	100.0	0	100.0
5	93.9	5	78.6	5	99.5	5	98.2	5	86.9
10	92.0	10	58.0	10	97.9	10	97.0	10	77.4
15	91.3	15	53.6	15	97.0	15	82.8	15	74.9
20	90.8	20	36.9	20	96.1	20	82.2	20	70.7
25	88.5	25	33.6	25	95.1	25	81.9	25	69.7
30	80.9	30	32.6	30	93.5	30	79.9	30	63.7

Halemba Mine

Material characteristics

The weight of coal grains submitted to crushing ranged from 7.8 kg to 21 kg and their sizes varied from 28 x 24 x 15 cm to 44 x 34 x 24 cm. The weight of stone grains ranged from 15.4 kg to 38.6 kg and their sizes varied from 40 x 21 x 16 cm to 66 x 24 x 14 cm.

Tests of dropping coal

The smallest number of discharges required to break up the coal grains completely was 5. In four trials, coal was broken up after 7 discharges. In seven individual trials coal was completely crushed to <80 mm after 5, 9, 12, 13, 14, 17 and 18 discharges. The average of the number of coal discharges during crushing was 9.9. The results of coal discharges are summarized in Table 4.

Table 4. Tests of dropping coal – Halemba Mine [12].

Sample 1 W_z=31x26x16 cm M_z=12.27 kg		Sample 2 W_z= 43x34x24 cm M_z= 21.00 kg		Sample 3 W_z= 30x20x18 cm M_z= 9.66 kg		Sample 4 W_z= 28x24x15 cm M_z= 7.66 kg		Sample 5 W_z= 34x28x18 cm M_z= 19.01 kg	
Number of discharges	Weight, %	Number of discharges	Weight, %	Number of discharges	Weight, %	Number of discharges	Weight, %	Number of discharges	Weight, %
0	100.0	0	100.0	0	100.0	0	100.0	0	100.0
5	-	5	25.2	5	19.4	5	11.5	5	30.8
		10	2.8	9	-	7	-	10	10.3
		12	-					15	2.5
								18	-
Sample 6 W_z= 31x25x20 cm M_z= 13.13 kg		Sample 7 W_z= 26x23x18 cm M_z=9.87 kg		Sample 8 W_z= 33x22x18 cm M_z= 11.48 kg		Sample 9 W_z= 35x22x14 cm M_z= 11.56 kg		Sample 10 W_z= 34x27x17 cm M_z= 11.72 kg	
Number of discharges	Weight, %	Number of discharges	Weight, %	Number of discharges	Weight, %	Number of discharges	Weight, %	Number of discharges	Weight, %
0	100.0	0	100.0	0	100.0	0	100.0	0	100.0
5	35.9	5	66.6	5	15.7	5	10.5	5	16.0
10	3.6	10	29.1	7	-	7	-	7	-
13	-	14	-						

Tests of dropping stone

In three trials, the stone was completely crushed to <80 mm after 14, 27, and 29 discharges. In two attempts, the stone was partly crushed. Weight loss of grain class> 80 mm was about 94%. The results are shown in Table 5.

Table 5. Tests of dropping stone - Halemba Mine [12].

Sample 1 W_z=34x32x23 cm M_z=19.90 kg		Sample 2 W_z= 66x17x11 cm M_z= 16.63 kg		Sample 3 W_z= 66x29x14 cm M_z= 38.61 kg		Sample 4 W_z= 40x21x16 cm M_z= 15.37 kg		Sample 5 W_z= 41x32x21 cm M_z= 31.12 kg	
Number of discharges	Weight, %	Number of discharges	Weight, %	Number of discharges	Weight, %	Number of discharges	Weight, %	Number of discharges	Weight, %
0	100.0	0	100.0	0	100.0	0	100.0	0	100.0
5	60.4	5	53.9	5	79.1	5	56.2	5	53.4
10	32.2	10	35.8	10	55.6	10	26.2	10	36.2
15	14.4	15	18.9	15	47.3	14		15	13.8
20	3.4	20	11.6	20	33.1			20	2.2
25	2.4	25	10.6	25	26.8			25	2.0
27	-	30	5.3	30	18.0			29	-

Pokój Mine

Material characteristics

The weight of coal grains submitted to crushing ranged from 4 kg to 21.4 kg and their sizes varied from 30 x 21 x 9 cm to 36 x 24 x 19 cm. The weight of stone grains ranged from 5.3 kg to 25.8 kg and their sizes varied from 24 x 24 x 7 cm to 44 x 22 x 19 cm.

Tests of dropping coal

The smallest number of discharges required to completely break up the coal grains was 7. In seven individual trials coal was completely crushed to <80 mm after 7, 9, 10, 12, 15, 27 and 29 discharges. In two trials, coal was broken up after 13 discharges. In a single sample, the grain was not completely crushed after 30 discharges, and 5.8% of the grains remained. The average number of coal discharges during crushing was 16.5. The results of coal discharges are summarized in Table 6.

Table 6. Attempts to dislodge coal – Pokój Mine [12].

Sample 1 W_z=57x21x19 cm M_z=16.31 kg		Sample 2 W_z= 25x25x20 cm M_z= 11.26 kg		Sample 3 W_z= 18x15x13 cm M_z= 3.23 kg		Sample 4 W_z= 26x20x19 cm M_z= 10.18 kg		Sample 5 W_z= 27x27x11 cm M_z= 5.89 kg	
Number of discharges	Weight, %	Number of discharges	Weight, %	Number of discharges	Weight, %	Number of discharges	Weight, %	Number of discharges	Weight, %
0	100.0	0	100.0	0	100.0	0	100.0	0	100.0
5	8.8	5	17.6	5	84.8	5	11.6	5	54.3
9	-	10	-	10	49.5	7	-	10	37.7
				15	43.0			15	19.0
				20	34.1			20	8.5
				25	11.8			25	6.6
				27	-			30	5.8

Sample 6 W_z= 42x25x19 cm M_z= 15.90 kg		Sample 7 W_z= 30x21x9 cm M_z=3.95 kg		Sample 8 W_z= 25x23x11 cm M_z= 4.03 kg		Sample 9 W_z= 36x34x19 cm M_z= 21.38 kg		Sample 10 W_z= 29x24x14 cm M_z= 8.18 kg	
Number of discharges	Weight, %	Number of discharges	Weight, %	Number of discharges	Weight, %	Number of discharges	Weight, %	Number of discharges	Weight, %
0	100.0	0	100.0	0	100.0	0	100.0	0	100.0
5	19.6	5	32.9	5	49.9	5	20.0	5	60.9
10	5.8	10	19.0	10	13.9	10	2.7	10	20.9
13	-	15	15.2	13	-	12	-	15	-
		20	13.4						
		25	11.9						
		29	-						

Tests of dropping stone

In two individual trials the stone was completely crushed to <80 mm after 10 and 17 discharges. In three trials, the stone was partly crushed. Weight loss of grain class> 80 mm was about 69%. The results are shown in Table 7.

Table 7. Attempts to dislodge the stone – Pokój Mine [12].

Sample 1 W_z=44x22x19 cm M_z=25.76 kg		Sample 2 W_z= 24x24x7 cm M_z= 5.31 kg		Sample 3 W_z= 26x16x8 cm M_z= 5.53 kg		Sample 4 W_z= 33x24x14 cm M_z= 20.32 kg		Sample 5 W_z= 23x23x18 cm M_z= 15.49 kg	
Number of discharges	Weight, %	Number of discharges	Weight, %	Number of discharges	Weight, %	Number of discharges	Weight, %	Number of discharges	Weight, %
0	100.0	0	100.0	0	100.0	0	100.0	0	100.0
5	93.0	5	22.4	5	36.9	5	96.5	5	80.0
10	65.8	10	-	10	19.9	10	90.0	10	43.5
15	58.5			15	18.6	15	87.3	15	29.1
20	55.4			17	-	20	77.1	20	15.9
25	39.9					25	71.4	25	12.1
30	29.3					30	69.2	30	4.7

3.3. Comparative analysis of test results

In order to compare the susceptibility to crushing, the class 80-0 mm output was analyzed as a function of the number of discharges for the tested run-of-mine. For each tested material (coal, stone), average values of discharges were calculated. The calculations included the weight of analyzed grains.

The results of the calculations are collected in Table 8 and illustrated graphically in Figures 5-7.

Table 8. Class 80-0 mm products depending on the number of discharges [12].

Number of discharges	Bielszowice Mine		Halemba mine		Pokój Mine	
	Coal, %	Stone, %	Coal, %	Stone, %	Coal, %	Stone, %
0	0	0	0	0	0	0
5	57.5	7.3	88.0	36.9	73.9	18.3
10	85.4	13.0	97.8	59.6	91.7	40.5
15	95.7	17.3	99.6	76.5	96.9	47.1
20	98.9	20.9	100.0	86.8	97.9	55.2
25	99.5	22.4	-	89.1	98.8	63.2
30	99.7	26.1	-	89.7	99.7	69.1

The greatest susceptibility to crushing was characterized by the run-of-mine from the Halemba Mine. The coal crushed completely after 18 discharges at the drop height of 2 m, and the average discharge of coal was 9.9.

After making 18 discharges of stone grains in the 80-0 mm class, about 85% of the material was crushed. A small difference in susceptibility to crushing, to the assumed grain size, between coal and stone can limit the effect of selective crushing of material in a drum crusher.

The run-of-mine from the Bielszowice Mine and the Pokój Mine was characterized by a smaller susceptibility to crushing. The run-of-mine was not crumbled completely.

At the maximum number of discharges at the level of 30 the percentage output of crushed coal in 80-0 mm class from the Bielszowice Mine and Pokój Mine was 99.7%.

The above results were obtained at the average number of coal discharges at the level of 16.8 (Bielszowice Mine) and 16.5 (Pokój Mine). After making 30 discharges of stone grains from the Bielszowice Mine in the 80-0 mm class, about 26.1% of the material was crushed, and from the Pokój Mine it was 69.1%.

The results of susceptibility tests to crushing stone >80 mm from the run-of-mine of the Pokój Mine and mainly of the Bielszowice Mine, despite the lower "brittleness" of coal grains, indicate a possibility of obtaining positive results of the material selective crushing in a drum crusher.

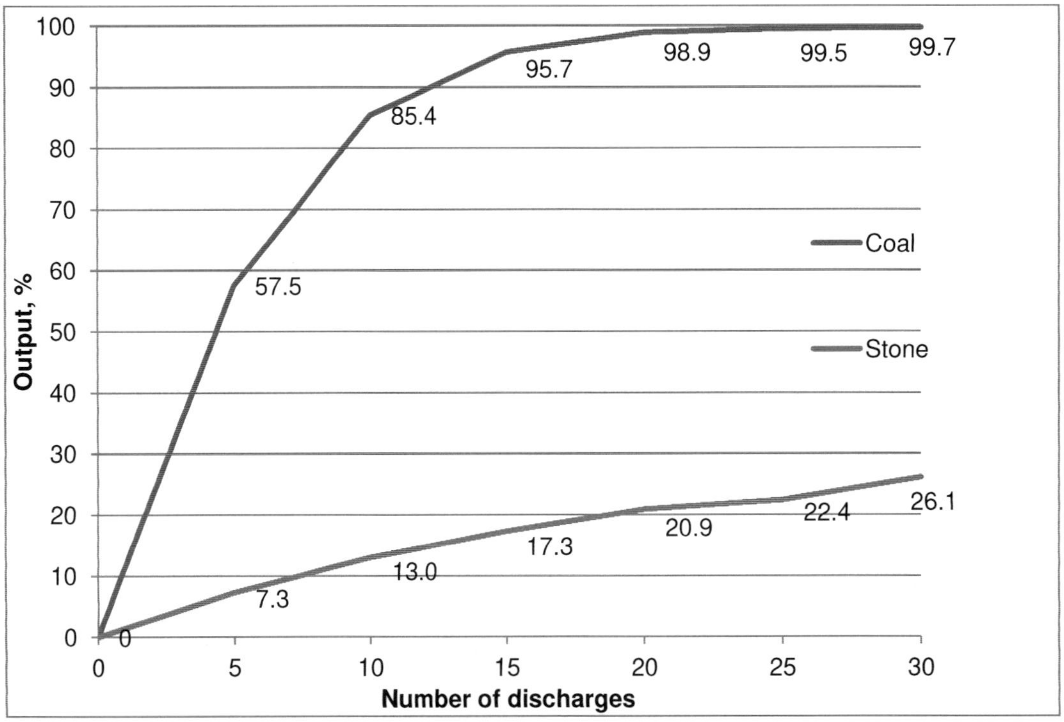

Figure 5. Class 80-0 mm output depending on the number of discharges - Bielszowice Mine [12].

IOP Conf. Series: Materials Science and Engineering **679** (2019) 012011 doi:10.1088/1757-899X/679/1/012011

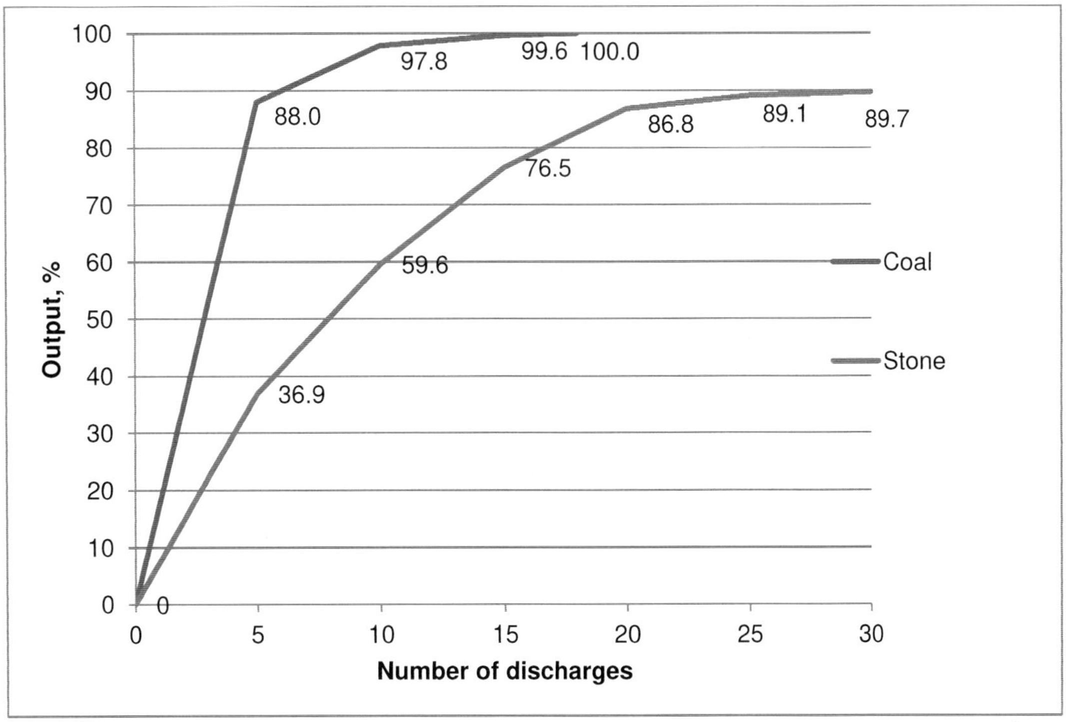

Figure 6. Class 80-0 mm output depending on the number of discharges – Halemba Mine [12].

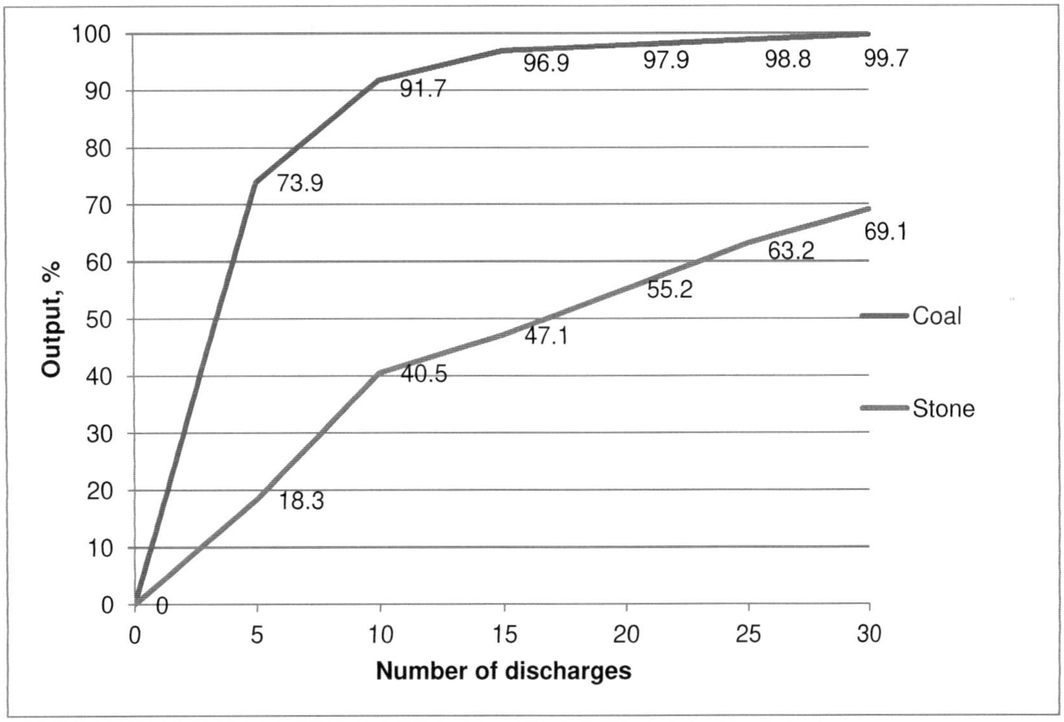

Figure 7. Class 80-0 mm output depending on the number of discharges – Pokój Mine [12].

IOP Conf. Series: Materials Science and Engineering **679** (2019) 012011 doi:10.1088/1757-899X/679/1/012011

4. Conclusions

The analysis of the test results, aimed at a determination of the possibility of using a selective crushing process for coking coal (type 34.2) and gas coal (type 33), showed that the use of drum crushers will enable partial deshaling of the run-of-mine with slight losses of coal in the waste.

During the tests, simulating the operation of the drum crusher with 80 mm holes, these losses were up to 4.3% after 15 discharges and up to 0.6% after 30 discharges.

Coal grains from the Halemba Mine were most susceptible to crushing. Simultaneously, a high susceptibility of gangue grains to crushing will reduce the selectivity of the process and will cause that a significant part of the crushed stone grains will participate in further technological processes. Depending on the number of discharges, their output, in relation to the total numbers of these grains in the feed, varied from 36.9% for 5 discharges to 89.7 for 30 discharges.

The biggest differences in susceptibility to crushing of coal and stone were recorded in the case of the run-of-mine from the Bielszowice Mine, which after 30 discharges exceeded 70%.

Despite of the fact that both types of the run-of-mines (Halemba Mine and Bielszowice Mine) represent coal 34, the obtained results show a significant difference in the susceptibility to crushing of the tested material.

Therefore, in order to determine a possibility of using the selective crushing process and ensure its high effectiveness, the decision to use a drum crusher should be preceded by tests enabling to determine the feed crushing potential.

References

[1] Doros Z and Osoba M Zwiększenie efektywności wydzielania gruboziarnistych odpadów kamiennych z urobku węgla surowego. *Zeszyty Naukowe. Górnictwo* z. **231**. Wydawnictwo Politechniki Śląskiej. Gliwice 1996

[2] Osoba M Odkamienianie urobku surowego węgla kamiennego. *Górnictwo i Geologia* 2011 Tom **6** Zeszyt 2 s. 167-179

[3] Review of the World Wide Status of Coal Preparation Technology, Report No. COAL R199DTI/Pub URN 00/1205, G Budge, J Brough and J Knight RJB Mining Consultancy Ltd and D Woodruff and L McNamara Baker Process, 2000

[4] Designing the Coal Preparation Plant of the Future, Barbara J. Arnold, Mark S. Klima, Peter J. Bethell, Society for Mining, Metallurgy, and Exploration; 1 edition, 2007

[5] https://www.mbe-cmt.com/index.php?id=48

[6] Jędo A Osadzarka z kołem odwadniającym do wzbogacania urobku węgla kamiennego. Mechaniczna przeróbka kopalin i gospodarka odpadami w aspekcie ochrony środowiska. materiały konferencyjne. Wydawnictwo CMG KOMAG. Szczyrk 1995

[7] Sorek S Eksploatacja złóż surowców mineralnych oraz przeróbka pozyskanego surowca w zamkniętym procesie wydobywczym. System wydzielania i lokowania odpadów w warunkach dołowych. Koncepcja instalacji do wydzielania odpadów z substancji palnej i lokowania ich w podziemiach kopalń. CMG KOMAG. Gliwice 2001 (not published)

[8] https://pdfs.semanticscholar.org/594c/1193bf93cef97a73f9006bf7162100c738d0.pdf

[9] http://jbindustrial.com/wp-content/uploads/2017/04/Pennsylvania-Crusher-Bradford-Breaker.pdf

[10] Gerus T and Wawrzyniak A Selektywne kruszenie jako metoda wydzielania odpadów z urobku surowego. *Maszyny Górnicze* 2000 R. 18 nr **1**. s. 40-45

[11] Kowol D et al. 2010 Poprawa i stabilizacja charakterystyki ilościowo – jakościowej urobku węglowego kierowanego do zakładu przeróbczego poprzez zastosowanie procesu odkamieniania w warunkach dołowych kopalni (not published)

[12] Kowol D et al. 2018: Wykonanie badań węgla w celu określenia wytycznych dla zabudowy kruszarki bębnowej typu Bradford oraz określenia jej parametrów technicznych dla KWK Ruda Bielszowice. Gliwice (not published)

IOP Conf. Series: Materials Science and Engineering **679** (2019) 012012 doi:10.1088/1757-899X/679/1/012012

Investigations into the wear rate of conical picks with abrasion-resistant coatings in laboratory conditions

K Krauze[1], Ł Bołoz[1]*, T Wydro[1] and K Mucha[1]

[1]AGH University of Science and Technology, Al. Mickiewicza 30, 30-059 Cracow, Poland

*E-mail: boloz@agh.edu.pl

Abstract: As part of own works performed by the Department of Machine Engineering and Transport of AGH University of Science and Technology, investigations into the rate of conical picks' wear were carried out. For this purpose, the previously developed methodology for running experiments at a specialist laboratory test stand was applied. Tests were carried out for a total of seven types of conical picks with the same geometric parameters, but made of various grades of steel and with the body working surface covered by abrasion-resistant materials. The conducted investigations and the analysis of results allowed recommending conical picks made of S235 steal and hardfaced with EP-TB-1-6 or EP-TB-2-46 electrodes. However, it seems advisable to continue attempts aimed at finding a protective coating for the body of the pick.

1. Indroduction

The works described in this article were carried out at the Department of Machinery Engineering and Transport of AGH University of Science and Technology, where the wear of conical picks was measured at a specialist laboratory test stand for testing the resistance of getting in the process of cutting or drilling. These tests are conducted in order to develop a methodology and choose materials to protect the body, and more specifically, the working surfaces having contact with mined rock against abrasive wear, in particular in abrasive minerals. Figure 1 contains a photograph of a conical pick the body of which has worn off and exposed the insert made of sintered carbide. Further use of this pick will result in the insert falling out as well as the wear and destruction of the pick. Therefore, providing protection for this part of the pick by subjecting it to thermal treatment or covering its surface with abrasion-resistant materials should increase the durability of these tools. It is noteworthy that currently one can find conical picks with bodies protected in the process of hardfacing with abrasion-resistant materials. However, the cost of these tools and their problematic quality cause that they are not competitors for the currently produced conical picks of alloy steels.

Conical picks are the subject of research and development works carried out in many centres all over the world. These investigations are mainly aimed at finding a solution that will ensure the highest resistance to abrasion. Tests were conducted for conical picks with bodies protected against abrasive wear by hardfacing, covering them with abrasion-resistant coatings or with rings made of sintered carbides [3, 10, 11, 12]. The works performed in order to explore the mechanism of conical picks' abrasion wear [4], wear prediction [5] or the possibility of supporting the cutting process [7] were conducted. Tests were carried out for sintered carbide [13] and for a complete conical pick [14] or for conical picks that make up the cutterhead [6].

Also, implementation works aimed at applying modern tools and machines in difficult conditions [1, 9] as well as alternative tools in the form of discs [8] were carried out. The works were also focused on the procedures of evaluating the quality of conical picks enabling the best offer to be selected by the user during public tenders [2].

Measurement of conical picks' wear rate is aimed at determining their durability. It must always be performed in the same conditions so that the results are credible, repeatable and probabilistic. This will allow for evaluating and comparing the conical picks' durability. In industrial conditions the durability of picks is usually defined as the ratio of the number of picks to the amount of output. It is the most frequently the number of picks used to obtain 1000 Mg or m^3 of output, whereas in laboratory conditions the wear rate (durability) of the pick or picks can be the most effectively determined by the loss of mass in relation to the volume of rock sample cut by this pick or picks.

Figure 1. Conical picks on a screw cutterhead: a) with an exposed insert made of sintered carbide, b) with a slipped insert of sintered carbide.

In this case, wear rate tests were carried out for conical picks made of various grades of alloy steels, having the same constructional parameters and working part surfaces protected by various hard materials in the process of hardfacing. As mentioned before, the investigations were conducted at a specialist test stand at the Department of Machinery Engineering and Transport of AGH University of Science and Technology.

2. Subject of research

Investigations into the rate of wear were conducted for seven types of conical picks made of various types of steel, commercial or hardfaced with different electrodes, which, in principle, increase their durability. All the picks had a working length of 90 mm and tip angle $2\beta u = 90°$, equipped with tips made of sintered carbide having a diameter $\phi 22$. The following picks were selected for tests:

- commercial picks made of 35HGS steel, thermally treated to reach the working part hardness of 54 HRC – 4 picks – marked as NS-01 (Figure 2a),
- picks made of 35HGS steel, where the working part surface was hardfaced with a Capilla HR MAG electrode by the TIG method – 4 picks – marked as NS-02 (Figure 2b),
- picks made of 35HGS steel, where the working part surface was plasma hardfaced with a Capilla HR MAG electrode – 4 picks – marked as NS-03 (Figure 2c),
- commercial picks made of S255 steel, thermally treated to reach the working part hardness of 45 HRC – 4 picks – marked as NS-04 (Figure 2d),
- picks made of S235 steel, manually hardfaced with an EP-TB-1-6 electrode having a hardness of approximately 50 HRC – 2 picks, and picks made of the same steel, hardfaced with an EP-TB-2-46 electrode having a hardness of ca 60 HRC – 2 picks – marked as NS-05 (Figure 2e),

- commercial picks made of 35HGS steel and thermally treated to reach the working part hardness of HRC 48 – 4 picks – marked as NS06 (Figure 2f),
- picks made of 35HGS steel, manually hardfaced with an EP-TB-1-6 electrode having a hardness of 50 HRC – 2 picks, and picks made of the same steel, hardfaced with an EP-TB-2-46 electrode having a hardness of 60 HRC – 4 picks – marked as NS-07 (Figure 2g).

Figure 2. Examples of picks selected for tests: a) commercial NS-01, b) hardfaced NS-02, c) hardfaced NS-03, d) commercial NS-04, e) hardfaced NS-05, f) commercial NS-06, g) hardfaced NS-07.

A total of twelve technically sharp commercial picks (NS-01, NS-04, NS-06) and 16 picks with the working body covered with abrasion-resistant coatings (NS-02, NS-03, NS-05, NS-07) were selected for tests. The NS-04 and NS-05 picks were made of S235 steel, whereas the NS-06 and NS-07 picks – of 35HGS steel. It should be noted that two picks made of S235 steel were coated with two different abrasion-resistant layers (NS-05, NS-07) so as to evaluate the adhesion of these layers to various grades of steel; during investigations they were positioned in the same holders.

3. Research plan and methodology

To assess the durability of picks provided for tests, the rate of their wear was assumed to be the ratio of the total loss of the picks' mass to the volume of output obtained in the process of artificial rock cutting. The adopted definition of the parameter determining the durability of picks as well as requirements regarding the rate of their wear resulted in the necessity of adopting the following research plan:

- preparation of a cement-sand sample (cement, sand, aggregate, water) having a required and empirically determined single-axis compressive strength,
- preparation of a test disc with appropriate holders,
- preparing and marking the picks to be tested as well as measuring their mass,
- mounting 4 conical picks on the test disc,
- cutting in laboratory conditions, with a constant value of feed rate and cutting,
- disassembly of picks and measurement of their mass,
- measurement of the volume of output obtained during the work of picks subjected to testing,
- calculation of C2 coefficient, determining the picks' wear rate.

Implementation of the adopted research plan and methodology required determining or adopting the following major parameters related to the process of cutting with the picks in question:

- cutting a cement-sand sample consisting of cement, sand and basalt aggregate with uniaxial compressive strength $R_c = 30 \div 35$ MPa,
- sample mass density $\gamma_w = 2.3$ Mg/m^3,
- volume of whole sample at the test stand $V_p = 8.125$ m^3,
- reference volume of the sample (for one set of picks) $V_w = 0.5$ m^3,
- advance speed: $v_{pu} = 0.05$ m/min,
- number of revolutions of the test disc: n = 42 rpm,
- length of the cut: 50 mm,
- web cut: 152 mm,
- diameter of the test disc with picks: ϕ1863 mm.

When the picks' mass before and after tests as well as the volume of output obtained with these picks have been determined, it is possible to determine the parameter characterizing the examined picks' wear rate. The rate of picks' wear (durability) should be determined from the following dependence.

$$C2 = \frac{\Delta m}{m} \cdot \frac{V_w}{V_u} \tag{1}$$

where:
C2 - picks' wear rate,
Δm - loss of pick's mass during tests (body and tip), g
m - picks' mass before tests, g
V_w - reference volume of the sample, m^3
V_u - volume of the sample cut during tests, m^3.

Investigations into all types of picks were conducted in accordance with the provided research plan and methodology. It should be noted that the lower the value of C2, the lower is the pick's wear. Laboratory tests are particularly recommended in this case, but they need to be carried out at a special test stand fulfilling the requirements of the adopted research methodology. For this reason, the investigations in question will be done at a test stand for testing the process of cutting with cutterheads possessed by KIMiT AGH (Department of Machinery Engineering and Transport of AGH University of Science and Technology).

4. Description of the test stand

The investigations were conducted at a laboratory test stand for testing the process of cutting or rotary drilling with single cutting tools or cutterheads. It is used for complex laboratory tests connected with the broadly understood process of rock cutting. The test stand enables particular cutterheads to cut an artificial or natural rock sample in laboratory conditions. The test stand consists of three main subassemblies (Figure 3):

- cutterhead drive assembly,
- sample mounting and driving assembly,

– measurement and control system.

The cutterhead drive assembly is based on a foundation and consists of an AC induction motor powered by a DTC indirect frequency converter having an output of 250 kW (supply voltage – 3x400V; number of revolutions - 1487 rpm). The torque from the motor shaft is transmitted onto the shaft of the cutterhead by a mechanical transmission gear with a transmission ratio $i = 28$.

Additionally, a torque meter has been mounted on the motor drive. The system configuration allows for the drive's work in two ranges, so-called constant torque control (up to 50 rpm) and constant power control (up to 120 rpm). At its end, the drive assembly is equipped with a spline, on which the cutterhead has been mounted.

A rock sample moves longwise and crosswise in relation to the cutterhead. Both the longitudinal and transverse movement of the rock sample is forced by hydraulic servomotors, which enable achieving a feed rate within a range of 0 to 9.9 m/min, feed force up to 150 kN, longitudinal travel up to 2.5 m, and transverse travel up to 1.3 m. The length of the rock sample reaches 2.5 m, the width 1.3 m and the height 2.5 m.

The test stand enables testing cutterheads with a maximum diameter of 2.2 m and a maximum web of 1.0 m. The output obtained in the process of cutting is provided into a container moving on guide bars together with the sample subjected to cutting. Such a solution enables collection of the output in a continuous way during the cutting process.

Figure 3. Laboratory test stand for investigating the process of cutting or rotary drilling with single cutting tools or cutterheads: a) operator's cab; b) concrete sample with its mounting and feed assembly as well as a test cutterhead; c) cutterhead drive assembly.

The system for controlling and recording particular values related to the cutting process is placed in the cab which reduces the level of noise and contamination. The control system was created using

PLC controllers. A touchpad and remote control are used to visualize and control the process. The operator can change all the process parameters and monitor the values being measured and recorded on a current basis. The measurement system consists of sensors that enable, either directly or indirectly, the following values to be measured and recorded:

- longitudinal feed rate,
- transverse feed rate,
- cutterhead's rotary speed,
- cutting resistance torque,
- longitudinal feed resistance force,
- transverse feed resistance force.

Additionally, the test stand provides the possibility of measuring dust during tests and analysing the grain size distribution of the obtained output.

5. Research and preparation of results

The translational motion of the rock sample and the rotary movement of the cutting element cause the destruction of the artificial rock's cohesion by cutting it with picks. Then it is possible to measure the mass loss of the working picks and the volume of output obtained after tests. Therefore, it is very important to provide constant and repeatable conditions of measurement throughout the picks' wear tests.

Before starting the tests, the following were done:

- inspection of the construction and operation of the test stand and drive,
- inspection of the measurement system,
- mounting the tested picks on the cutting element by means of an appropriate adapter (Figure 4),
- setting the feed rate, cutterhead revolutions and web.

Tests were done for seven sets of picks. The picks were prepared in a way enabling their direct assembly on the test disc. The investigations were carried out in accordance with the established research plan.

The picks on the test disc formed a pick system, the diagram of which has been presented in Figure 5. The picks were located on the circumference, at 90° intervals. Each set of picks was mounted in holders installed on the disc, which were numbered 1 to 4. Upon completion of the cutting process, the picks were photographed (Figure 6) and their mass was measured.

Figure 4. Examples of conical picks prepared for tests.

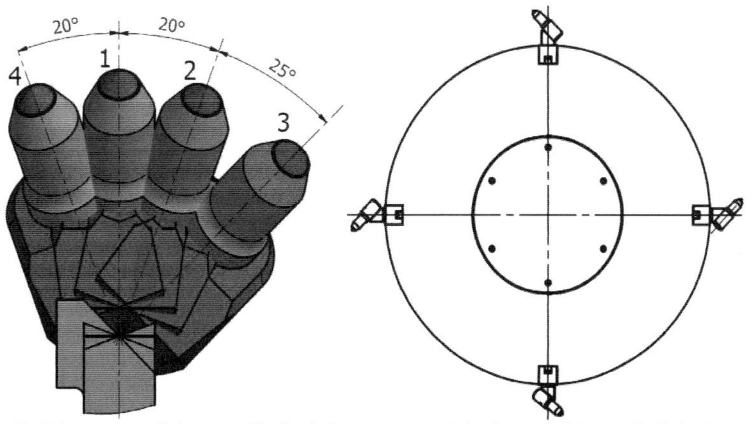

Figure 5. Diagram of the applied pick system with the marking of picks' numbers.

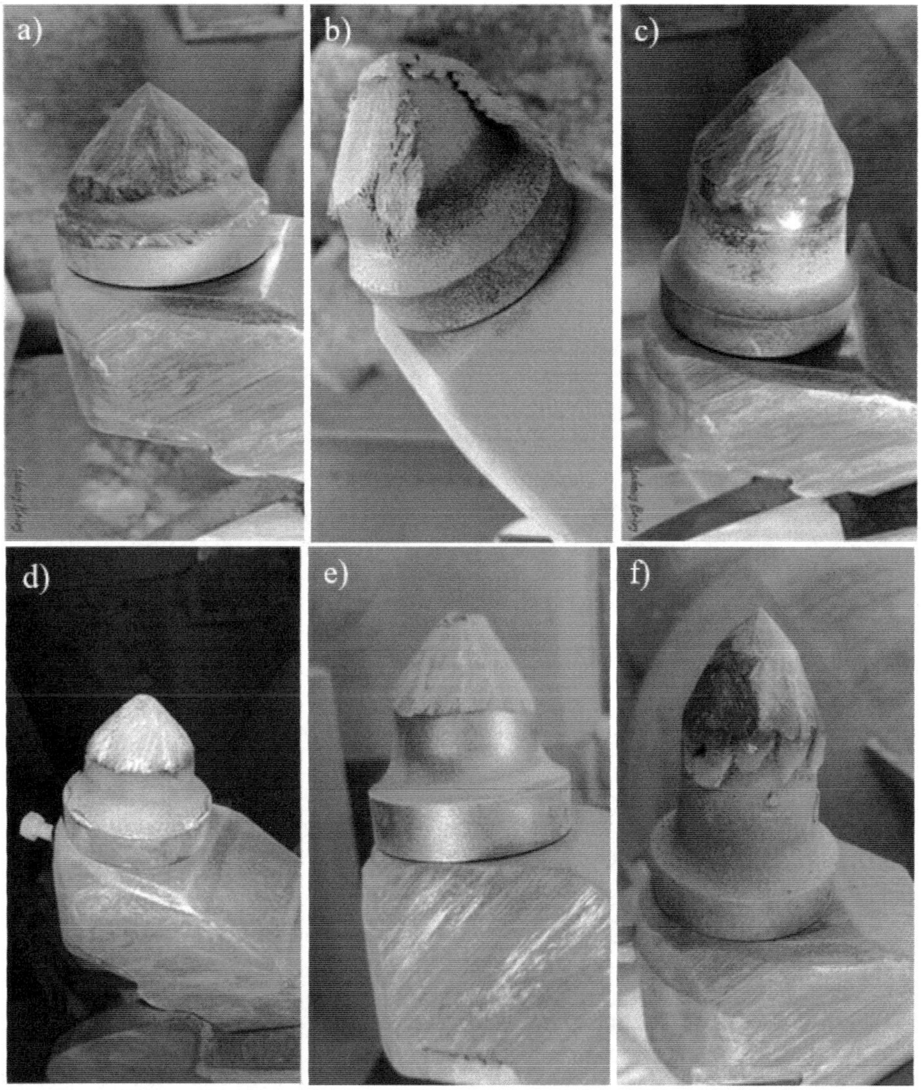

Figure 6. Selected picks after tests: a) commercial NS-01, b) hardfaced NS-02, c) hardfaced NS-03, d) commercial NS-04, e) commercial NS-05, f) hardfaced NS-06.

IOP Conf. Series: Materials Science and Engineering **679** (2019) 012012 doi:10.1088/1757-899X/679/1/012012

Next, the volume of output V_u, which was different for each sample (each type of pick) was determined. Based on the obtained values, parameter C2 defining the picks' wear rate was calculated, in accordance with the adopted research plan and methodology. Examples of test results for NS-01 commercial picks have been given in Table 1.

Table 1. Test results and parameters determined for NS-01commercial picks.

NS-01 commercial picks							**Reference sample volume V_w = 0.5 m^3**
No.	**Position of pick on the test disc**	**Producer marking**	**Pick's mass**		**Pick's mass loss $\Delta m = m - m_p$ [g]**	**Volume of output V_u [m^3]**	**C2 [–]**
			before test m [g]	**after test m_p [g]**			
1	1	–	1855.12	1388.06	467.06	0.103	1.23
2	2	–	1843.75	1249.77	593.98		1.57
3	3	–	1845.04	1374.34	470.70		1.24
4	4	–	1844.09	1356.70	487.39		1.29
Mean value of C2 parameter of the pick's relative mass loss							**C2: 1.33**

6. Research results analysis

The conducted tests and the obtained results enable evaluating the durability of particular picks and the influence of protective coatings on this durability. It should be clearly emphasized that the picks in question always cut in the same conditions (cutting and feed rate, pick system, rock sample), according to the research plan and methodology, and the measure of the rate of individual and average wear is C2. The value of this parameter indicates the pick's wear rate, and the lower this parameter is, the higher is the pick's durability.

The lowest mean value of C2 was noted for NS-05 picks (C2 = 0.41), and the highest for NS-02 picks (C2 = 1.48). In the first case, the picks had a protective coating on the working part, which was made in the process of hardfacing with an EP-TB-2-46 electrode, whereas in the second case – with a Capilla HR MAG electrode (hardfacing by the TIG method).

The protective coating does not protect the pick if it is not suited to the conditions of its work. Picks without a protective layer (NS-01, NS-04, NS-06), made of alloy steel and subjected to thermal treatment, have comparable durability. Therefore, NS-05 picks, followed by NS-03 and NS-07 are recommended for work with abrasive rocks (Table 2).

A similar trend can be observed when considering the individual C2 parameter and the position of the pick on the test disc. In this case, however, the assessment is similar, though less unambiguous. It is clearly visible – as could be expected – that the wear rate strongly depends on the position of the pick, i.e., the cutting resistance (the order of picks entering the rock mass).

Table 2. Mean C2 values and research observations – summary table.

List of picks – in total			
No.	**Pick**	**C2**	**Remarks**
1	NS-01 commercial picks	1.33	three picks worn asymmetrically, one pick worn symmetrically, all picks without carbides
2	NS-02 hardfaced picks	1.48	one pick worn asymmetrically, three picks worn symmetrically, all picks without carbides
3	NS-03 hardfaced picks	0.53	two picks worn asymmetrically; two picks worn symmetrically, only one pick without carbide
4	NS-04 commercial picks	1.36	four picks worn symmetrically, all picks without carbides
5	NS-05 hardfaced picks	0.41	four picks worn symmetrically, all picks with carbides
6	NS-06 commercial picks	1.08	one pick worn asymmetrically, three picks worn symmetrically, all picks without carbides
7	NS-07 hardfaced picks	0.76	four picks worn symmetrically, two picks without carbides

7. Summary

The conducted investigations into the rate of picks' wear in the process of artificial sample cutting indicate that their durability is strongly influenced by the type of protective coating on the pick's working part, the grade of steel used to produce the picks' bodies as well as the position of the pick on the cutterhead. It is therefore suggested that further searches for a protective layer to be applied on the pick's working part should be undertaken so as to establish which protective coating is the cheapest and the best for increasing picks' durability, in particular in hardly accessible and/or abrasive rocks.

References

[1] Bołoz Ł and Krauze K (2018) Ability to mill rocks in open-pit mining. In: 18th International Multidisciplinary Scientific Geoconference, Exploration and Mining, SGEM2018, Albena, Bulgaria, Voulme **2**, pp. 41-48.

[2] Bołoz Ł (2018) „Results of a study on the quality of conical picks for public procurement purposes", in Proceedings of the international conference on Human safety in work environment: operating machinery and equipment: integrated management systems: quality - environment - safety, 23-27 october 2018, Gdańsk-Nynashamn-Sztokholm-Tallin-Sztokholm-Nynashamn-Gdańsk, pp. 687-693.

[3] Chang, Soo-Ho, Lee Chulho, Kang Tae-Ho, Ha Taewook, Choi, Soon-Wook (2017) Effect of hardfacing on wear reduction of pick cutters under mixed rock conditions. *Geomechanics and engineering*, **13(1)**, pp. 141-159.

[4] Dewangan, Saurabh, Chattopadhyaya, Somnath (2015) Critical Analysis of Wear Mechanisms in Cemented Carbide. *Journal of materials engineering and performance*, **24(7)**, pp. 2628-2636.

[5] Gajewski J, Jedlinski L and Jonak J. (2013) Classification of wear level of mining tools with the use of fuzzy neural Network. *Tunnelling and underground space technology*, **35**, pp. 30-36.

[6] Gospodarczyk P, Kotwica K and Stopka G (2013) A new generation mining head with disc tool of complex trajectory, *Archives of Mining Sciences*, **58(4)**, pp. 985–1006.

[7] Kotwica K (2011) The influence of water assistance on the character and degree of wear of cutting tools applied in roadheaders. *Archives of Mining Sciences*, **5(3)**, pp. 353-374.

IOP Conf. Series: Materials Science and Engineering **679** (2019) 012012 doi:10.1088/1757-899X/679/1/012012

[8] Kotwica K (2018) Atypical and innovative tool, holder and mining head designed for roadheaders used to tunnel and gallery drilling in hard rock. *Tunnelling and Underground Space Technology*, **82**, pp. 493-503.

[9] Krauze K and Bołoz Ł (2018) Disc unit dedicated to mine abrasive rocks and in particular copper ores. In: 18th International Multidisciplinary Scientific Geoconference, Exploration and Mining, SGEM2018, Albena, Bulgaria, Voulme **2**, pp. 311-318.

[10] Krauze K, Bołoz Ł and Wydro T (2015) Parametric factors for the tangential-rotary picks quality assessment, *Archives of Mining Sciences*, **60(1)**, pp. 265-281.

[11] Krauze K, Bołoz Ł, Wydro T and Mucha K (2017) Durability testing of tangential-rotary picks made of different materials. *Mining – Informatics, Automation and Electrical Engineering*, no. **1**, pp. 26-34.

[12] Krauze K, Skowronek T and Mucha K (2016) Influence of the hard - faced layer welded on tangential – rotary pick operational part on to its wear rate. Archives of Mining Sciences 61(4), pp. 779 – 792.

[13] Nahak, Sakuntala, Dewangan, Saurabh, Chattopadhyaya and Somnath (2015). Discussion on Wear Phenomena in Cemented Carbide. In: Global Challenges, Policy Framework & Sustainable Development for Mining of Mineral and Fossil Energy Resources. Dhanbad: GCPF, **11**, pp. 284-293.

[14] Songyong L, Huifu; J and Xiaohui L (2017) Experimental research on wear of conical pick interacting with coal-rock. *Engineering failure analysis*, **74**, pp. 172-187.

IOP Conf. Series: Materials Science and Engineering **679** (2019) 012013 doi:10.1088/1757-899X/679/1/012013

Tests of flame-proof properties of supply-control system for a mining floor-loader

K Lesiak

KOMAG Institute of Mining Technology, Laboratory of Applied Tests, Pszczyńska 37, 44-101 Gliwice, Poland

E-mail: klesiak@komag.eu

Abstract. Tests of flame-proof properties of supply-control box for a floor-loader operating in underground mine workings, where methane and/or flammable dust explosion hazard occurs are presented. The tests included, among others, a determination of the explosion pressure, the enclosure strength to the pressure and the test of non-transfer of internal explosion. The tests enabled to localized the components, which required some improvements regarding their mechanical strength but also confirmed safety of the final design. The enclosure of the supply-control equipment of a floor-loader is designed within the HYDKOM 75 project co-financed by the National Centre for Research and Development.

1. Introduction

Floor-loaders are mobile machines installed on a caterpillar chassis, equipped with electrohydraulic drive, in which all the control functions are realized hydraulically. They serve, among others, for removal from the floor the waste rock generated at driving the roadway workings by means of blasting technique, for pulling materials and ripping the floor [1, 2, 3]. The hydraulic pump is driven by an electric motor. Due to the fact that the motor is supplied from an unwinding cable connected to the mine power network, the equipment mobility is limited and the supply cable is subjected to damages [2, 4]. A floor-loader with a battery drive, developed within the project entitled: "Innovative mobile machine with an universal electric drive system increasing the technical safety level" HYDKOM 75 within the Operational Programme Intelligent Development 2014–2020 eliminates all the above mentioned limitations. The project objectives include designing, manufacture and testing the battery driven floor loader to be operated in mine workings where a potential explosion hazards of methane and/or coal dust occurs.

The article presents the problems of the tests aimed at confirmation of a safe use of the state-of-the-art, innovative floor-loader, designed at KOMAG, in the workings of methane and/or coal dust explosion hazard.

2. "d" flameproof enclosure

The "d" flameproof enclosure is one of the methods for protecting the equipment to be operated in the areas, where an explosion hazard occurs potentially. This is the enclosure in which the components which may cause ignition of flammable gases are installed and which can withstand the pressure generated during an internal explosion of the explosive mixture and which protects against transfer of the explosion to the potentially explosive atmosphere surrounding the enclosure [5].

IOP Conf. Series: Materials Science and Engineering **679** (2019) 012013 doi:10.1088/1757-899X/679/1/012013

So-called flameproof connector or fire path is a characteristic component of such flameproof enclosures. This is a place, where two surfaces of enclosure components are contacted or where two enclosures are connected to eliminate transfer of internal explosion to the potentially explosive atmosphere [5]. Flange, cylindrical, flange-cylindrical as well as threaded connectors are most often used types of flameproof connectors. Examples of the selected connectors are presented in Figure 1, where L is a length of flameproof connector in the case a) and b) and in the case c), *L* is a sum of *c* and *d* assuming that $f \leq 1$ mm.

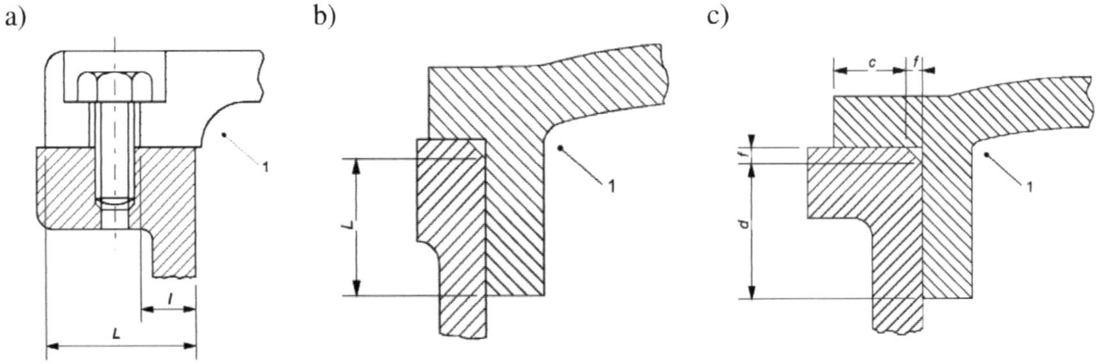

Figure 1. Examples of flameproof connectors a) flange, b) cylindrical, c) cylindrical- flange; 1 – internal enclosures, L – length of flameproof connector [5].

Seals made of compressive material are used sometimes to obtain the required IP protection level. In this case the seal is not taken into consideration in determining the length of flameproof connection and it cannot divide it. The examples of determination of flameproof connection in the case of using the seals are presented in Figure 2.

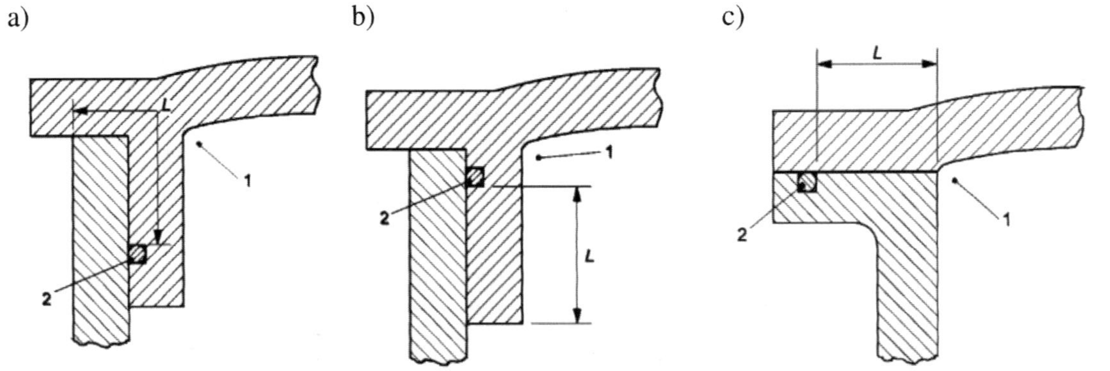

Figure 2. Examples of determination of flameproof connections in the case of using the seals; 1 – enclosure interior, 2 – seal, L – length of flameproof connection [5].

The flameproof connection has to extinguish a flame between the flameproof enclosure and the surrounding atmosphere. A flammable fuel, oxygen and high temperature are indispensable for ignition and sustaining the flame/explosion propagation. Elimination of one of these three components causes the flame extinguishing or lack of ignition possibility. Flameproof connection due to its proper design (e.g. length and roughness) causes reduction of temperature of a flame generated inside the flame enclosure so it can be distinguished. In the case of non-threaded connections, due to the elasticity of the screws connecting the two surfaces of the flameproof connection, during explosion inside the enclosure, the clearance between the enclosure surfaces increases slightly. Therefore, in the resulting clearance, gases formed during explosion / combustion are released and the pressure inside the enclosure drops.

3. Testing object and place of testing

The two-chamber flameproof enclosure of the supply-control module, designed and manufacture within the HYDKOM 75 project was the testing object (Figure 3). In the designed, battery powered floor-loader, "d" flameproof enclosure consisting of two separated flameproof chambers was used. One of the chambers had supplying cells with the Battery Management System (BMS), and in the second compartment there was a control system consisting of cells charger, power-electronic transducer or control-switching equipment.

All the tests of the flameproof box were conducted on the special test stand in the KOMAG Laboratory of Applied Tests.

During testing the flameproof, the box chambers were filled with the components simulating the final components.

Figure 3. Tested enclosure for the power supply-control module.

4. Testing the flameproof box

4.1. Test objective

Verification of design assumptions of the developed box for the supply-control module as well as a possibility of its use in the underground mine workings of the hard coal mining plants was the tests objective.

4.2. Scope of tests

Scope of the flameproof box verification tests regarding its flameproof ability included the following tests:

- Checking the flameproof connectors for conformity with the technical documentation and with the requirements of the PN-EN 60079-1:2014-12 Standard (before and after the pressure strength tests),
- Determination of the explosion pressure,
- Tests with overpressure,
- Tests for not transferring the internal explosion.

4.3. Testing procedure and the test results

4.3.1. Determination of a reference pressure. Determination of the reference pressure consists in determination of the maximum pressure that arises in the tested flameproof enclosure / chamber during

IOP Conf. Series: Materials Science and Engineering **679** (2019) 012013 doi:10.1088/1757-899X/679/1/012013

the forced explosion of the adequate explosive mixture. Adequate explosive mixture is a gas / gas concentration in the air that is characteristic for a given explosion group. For the first explosion group, which concerns mining plants threatened by the presence of methane and or combustible dust, the test mixture used to determine the reference pressure is a mixture of methane (9.8 ± 0.5)% in the air. When determining the reference pressure, three tests were performed for each chamber.

According to the project assumptions, the flameproof enclosure was equipped with O-ring seals or with glued rubber cord on all flameproof cylindrical and flange connections. Due to the above, the determination of the reference pressure was carried out with a set of seals installed. Assuming that the tested enclosure is intended for operation at temperatures above -20° C, tests were carried out at a pressure of the mixture equal to atmospheric pressure. The arrangement of pressure transducers and sources of ignition is presented in Figure 4, while Table 1 presents the recorded maximum reference pressures. The highest measured pressure for each chamber is the reference pressure.

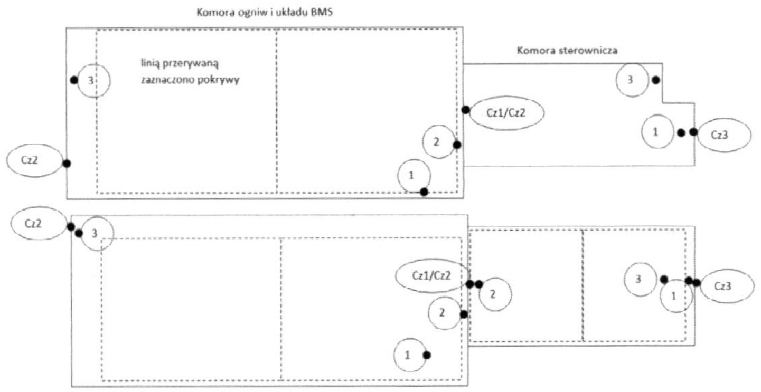

Figure 4. Arrangement of the measuring instruments during determination of the reference pressure (top and front view of the tested enclosure).

Table 1. Results of the reference pressure measurements.

Item	Type of the tested chamber	Ignition source	Pressure [bar]		
			Cz1	Cz2	Cz3
1	Power cells chamber	1	4.65	**6.03**	-
2		2	5.93	4.49	-
3		3	3.98	5.55	-
4	Control chamber	1	-	4.40	4.54
5		2	-	4.53	**4.64**
6		3	-	4.28	4.40

Pressure curve for the cells chamber for the ignition source No. 1 is presented in Figure 5.

IOP Conf. Series: Materials Science and Engineering **679** (2019) 012013 doi:10.1088/1757-899X/679/1/012013

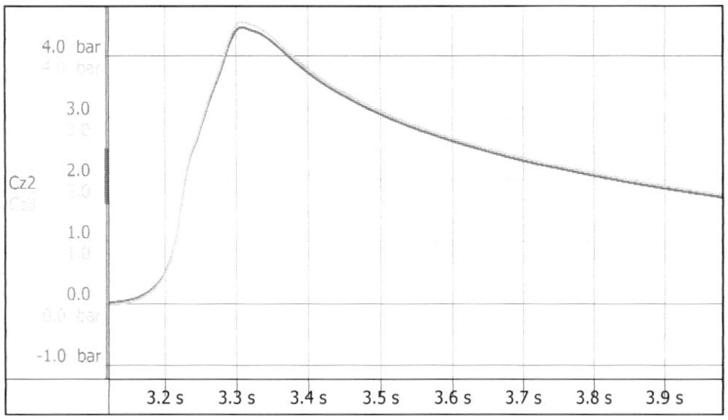

Figure 5. Sample pressure curve recorded during determination of the reference pressure.

4.3.2. Test with overpressure. The test is aimed at checking the resistance of the tested enclosure to pressure. The test may be carried out by one of two equivalent methods, static or dynamic. In the static method, for safety reasons, it is recommended to use an incompressible liquid. The test chamber is filled with a liquid at the appropriate pressure which is kept for at least 10 s. The dynamic method uses an explosive mixture which is ignited to obtain the required pressure 1.5 times the reference pressure during the explosion.

The tested enclosure obtains a positive result if, in a result of the overpressure test, none of the enclosure components is permanently deformed or damaged changing the protection level, and the flameproof connections are not permanently enlarged.

After testing the control box under overpressure, a positive result was obtained, confirming the required strength of this chamber. In the case of the cells chamber, both front covers were permanently deformed (bulged). Design assumptions included, among others, strictly defined external dimensions of the entire device (enclosure) and the number of power cells to be installed in the box. Meeting the design requirements forced, among others the thickness and profile of the covers, which were deformed, and thus affected their strength.

The above requirements were translated into the thickness of the designed covers and thus their strength.

As part of the improvements, the material from which covers were made, instead of S355JR, the new covers were made of S690QL steel. The new steel has a yield strength of ReH ≥ 690 MPa compared to 355 MPa for S355JR steel.

Due to the introduced changes, after making and installing new covers in the cells chamber, it was necessary to re-determine the reference pressure for this part of the cover. The results of the repeated tests are shown in the Table 2.

Table 2. Results of the reference pressure measurements in the case of new covers for the cells chamber.

Item	Type of the tested chamber	Ignition source	Pressure [bar]		
			Cz1	Cz2	Cz3
1	Power cells chamber	1	4.99	7.7	-
2		2	5.01	**8.6**	-
3		3	6.97	4.17	-

Higher explosion pressures were recorded in relation to the first tests.

The pressure resistance test for the cells chamber, with new installed covers, was performed by a static method. Due to experience with previous covers, apart of checking the flatness before starting the test, during the tests it was decided to install an additional displacement transducer. The transducer was installed in such a way that it was possible to measure the bulge in the midpoint of a larger front cover. The displacement measurements (l) were presented on one time axis with the pressure (p) exerted on the internal surfaces (forced inside) of the cells chamber. The time processes recorded in this way are presented in the form of a graph l = f (p) and presented in Figure 6.

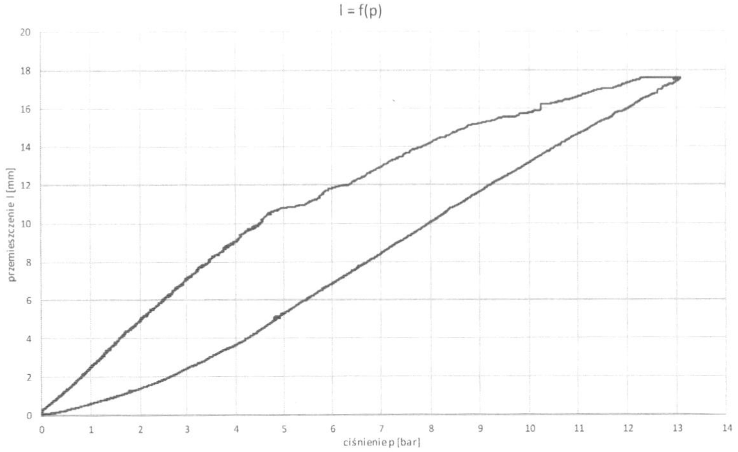

Figure 6. Characteristics l=f(p) recorded in a central point of the biggest cover of the cells chamber during the overpressure test.

After the test, no permanent deformation of any component of the checked chamber was found (this applies to both new covers and the rest of the structure being checked). No permanent enlargement of flameproof connections was found, which ultimately resulted in a positive result for the cells chamber.

4.3.3. Test for not transferring the internal ignition. The test for not transferring the internal ignition consists in placing the tested cover in the test chamber and filling both the tested enclosure and the test chamber with an explosive mixture. Then the mixture is ignited inside the tested enclosure and it is checked whether the explosion from the tested enclosure is not transferred to the test chamber. As in the case of determining the reference pressure, also in this case, the explosive mixture depends, among others from the explosive group to which the tested enclosure is intended.

Preparation of the object for tests includes removal of all gaskets, removal of excess grease protecting flameproof connections and obtaining the flameproof connections clearance equal to at least 90% of the maximum structural clearance. Mixture of hydrogen and methane [(58±1)% and (42±1)% hydrogen] in amount of (12.5±0.5)% in air was the test mixtures used during the tests.

After 10 tests of the cells chamber and 5 tests of the control chamber, there was no transfer of internal ignition for any of the chambers of the tested enclosure, what means obtaining the positive result of this test.

5. Conclusions

Tests of various types of flameproof enclosure designed at the KOMAG Institute of Mining Technology within the HYDKOM 75 project was discussed. In a result of the research work, it was necessary to introduce constructional corrections related to the mechanical strength of the designed flameproof enclosure. The effectiveness of the introduced changes was confirmed by the repeated verification tests.

IOP Conf. Series: Materials Science and Engineering **679** (2019) 012013 doi:10.1088/1757-899X/679/1/012013

The tests carried out in the Laboratory for Applied Tests finally confirmed the safety of the flameproof enclosure designed at the KOMAG Institute of Mining Technology and manufactured at Hydrotech S. A. within the HYDKOM 75 project for a set of power cells and the control part of the new floor-loader.

References

[1] Kalita M Spągoładowarka górnicza jako wielofunkcyjna maszyna robocza z podwoziem gąsienicowym o szerokości 1000 mm, Nowoczesne metody eksploatacji węgla i skał zwięzłych, TUR 2011, str. 270 -278, 2011

[2] Deja P, Kapuściński D, Niedworok A and Polnik B. Badania innowacyjnego układu zasilająco-sterującego spągoładowarki górniczej. *Maszyny Elektryczne: zeszyty problemowe* 2019 nr **2**(**122**) s. 15-19

[3] Kapuściński D and Polnik B Innowacyjny układ zasilająco-sterujący spągoładowarki górniczej. *Maszyny Elektryczne – Zeszyty Problemowe* 2019 nr **2**(**118**) s. 129-135

[4] Konsek R and Deja P Nowe rozwiązanie układu zasilająco-sterującego górniczej maszyny mobilnej ITG KOMAG *Maszyny Górnicze* 2018 nr **4** s. 67-77

[5] European standard PN-EN 60079-1:2014-12 Explosive atmospheres - Part 1: Equipment protection by flameproof enclosures "d"

IOP Conf. Series: Materials Science and Engineering **679** (2019) 012014 doi:10.1088/1757-899X/679/1/012014

Analysis of the process of loosening the rocks with different strength properties using the undercutting bolts

M Siegmund[1]* and J Jonak[2]

[1]KOMAG Institute of Mining Technology, Division of Machinery and Equipment, Pszczyńska 37, 44-101 Gliwice, Poland

[2]Lublin University of Technology, Faculty of Mechanical Engineering, Nadbystrzycka 36, 20-618 Lublin, Poland

*E-mail: msiegmund@komag.eu

Abstract: The process and the test results of the solid rock loosening by use of the fixed undercutting bolt are described. The tests were carried out within the RODEST project, OPUS 10 competition No. 2015/19/B/ST10/02817, financed by the National Science Centre, where a series of numerical modelling as well as laboratory and in-situ tests were planned. The test stand equipment as well as methodology for conducting the in-situ tests are presented. The tests were conducted in four mines. In each mine tens of solid rock loosening tests were conducted, during which the process parameters were recorded. Additionally, the loosened area outline in each successful test was recreated using 3D scanner. Data collected during in-situ tests as well as strength properties of the tested rocks enabled analysing the loosening process (forces, ranges) in relation to the rock type.

1. Introduction

According to the latest EU recommendations, it is advisable to search for alternatives to the known rock mining methods (mechanical and blasting), which is caused by the need to reduce the emission of gases harmful to human health and the environment, and the restrictions in using the mechanical mining technology in the area of critical buildings and industrial installations (due to e.g. generated vibrations), etc. Also extremely difficult mining and geological conditions as well as space and time limitations during rescue operations, practically eliminate the possibility of using conventional methods for development of mine workings [1, 2]. In such situations, the mining (loosening) processes are not expected to be carried out with the comparable operational and output parameters, but it is expected to ensure safety or even the possibility of carrying out the rocks loosening process.

The technology of rock fragmentation using bolts anchored in a solid rock, patented by KOMAG in several versions, is an alternative method to the traditional ones [3]. This method is safe because it does not create a risk of throwing out the mined rocks and can be used regardless of the level of methane hazard. The technology does not cause the rock mass fragmentation outside the strictly defined zone and in no way affects the immediate environment (no gas emissions or generated vibrations).

This solution has been developed and patented in several versions, mainly differing in the way of exerting tensile stress in the bolt. The main idea of the solution is shown in Figure 1. Lower energy consumption, while maintaining proper fixation of the element securing the bolt, is obtained by introducing tensile and shear stresses, whose limit values in the case of rocks are several times lower than the limit compressive stresses [1, 3, 4] . The method can be used for both solid and easy-to-be-mined rocks. The drilling direction can be any at will (vertical, horizontal, oblique). The method in its simplest form (made manually) does not guarantee much progress, however, it enables drilling of the roadway in any mining and geological conditions and is safe.

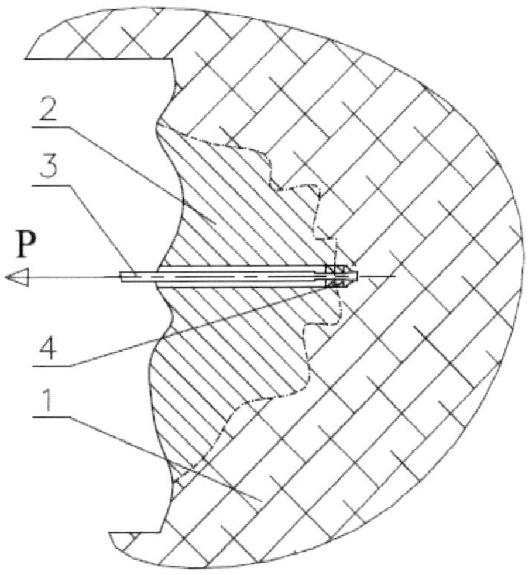

Figure 1. The concept of cutting the solid rock through destruction of its integrity: 1-solid rock, 2-loosened rock, 3-pulling rod, 4-expansion sleeve [1, 3, 4].

Research work on the technology of drilling the rescue tunnels by the method of destroying the rock integrity have been carried out at KOMAG for several years [4]. The first concept of the method of mechanical loosening of rocks and tests aimed at assessing the possibility of its application were conducted as a part of the INREQ project [5]. The state of the art in the field of bolts applications mainly applies to the fixation systems used in the assembly of industrial infrastructure in reinforced concrete engineering facilities. There is a wealth of knowledge in both experimental, empirical and FEM numerical studies on various aspects related to the formation of the bolt pull-out force, as well as the bolt assemblies and range - propagation of the crack path (loosening gap). European, Japanese and American standards, which allow to simplify the estimation of the pulling force in the context of the useful load capacity of the bolt are the result of this research work. Generally, for the purpose of calculating the bolting capacity, the simplified rock loosening models are adopted, i.e. detachment of the solid in the form of a cone or pyramid [6, 7, 8, 9, 10] (Figure 2).

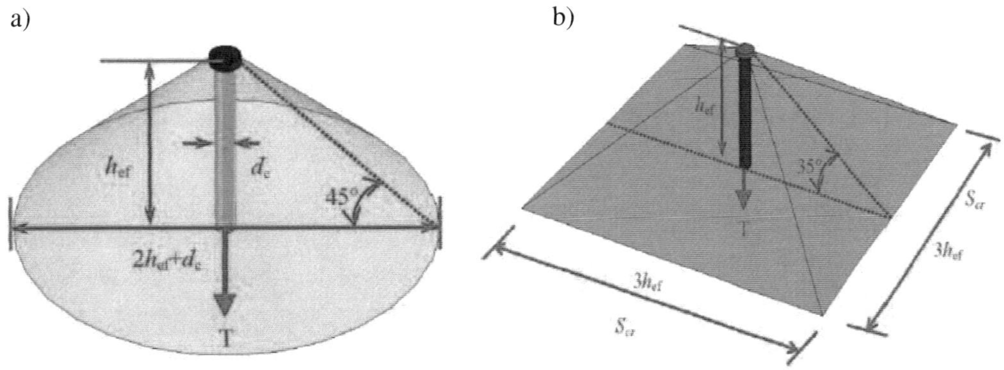

Figure 2. Simplified rock loosening models recommended for the practical use: a) conical, b) four-sided pyramid [11].

In the context of rocks loosening range, the practice shows that these are very large simplifications. The angles at the base of the cone, in practice are often more than 2 times smaller than the standard 35° or 45°, what means that the estimated ranges of loosening, and thus, e.g. the volume of the detached solids are significantly underestimated. The actual ranges of loosening (gap propagation) both in the light of literature analysis and own research work are shown in Figure 3.

a)

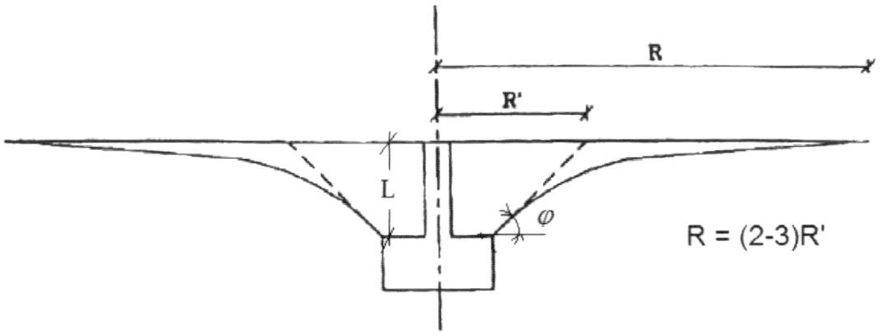

b)

Figure 3. Real propagation of the loosening path: a) concrete in the light of literature data [12], b) sandstone, in the light of own tests [13].

The knowledge resources existing in this area and the models used are not adequate to map phenomena corresponding to mining by loosening the rock fragments with a fixed bolt. Also the computer simulation tests carried out so far using the finite element method do not give an unequivocal answer regarding the scope of propagation of the loosening gap. It is necessary to compare different methods of crack propagation with in-situ tests to check their accuracy [14]. It is indispensable to carry out tests on shaping the loosening force depending on so called effective loosening depth (depth of undercut in the hole) as well as strength properties of the rock to determine possibilities of loosening the rocks with use of undercutting bolts. In order to fully understand the mechanism of loosening and strain in the pulled out rock, the RODEST project entitled: "Research and modelling of the mechanism for destruction of rock materials in the spatial state of shear and tensile stress" has been started. The project is realized by the scientific consortium: KOMAG Institute of Mining Technology together with the Lublin University of Technology and is financed by the National Science Centre as part of the OPUS 10 competition (project No. 2015/19/B/ST10/02817).

2. Testing the process of loosening the rocks using the undercutting bolts

2.1. Undercutting bolt

The stress in rock material during mining tests using a bolt can be referred to the problem of the load-bearing capacity of mechanical bolts. In the fixation technique, many types of fixing bolts are used, differing in the way they are fixed or loaded. The most commonly used types of mechanical bolts are shown in Figure 4.

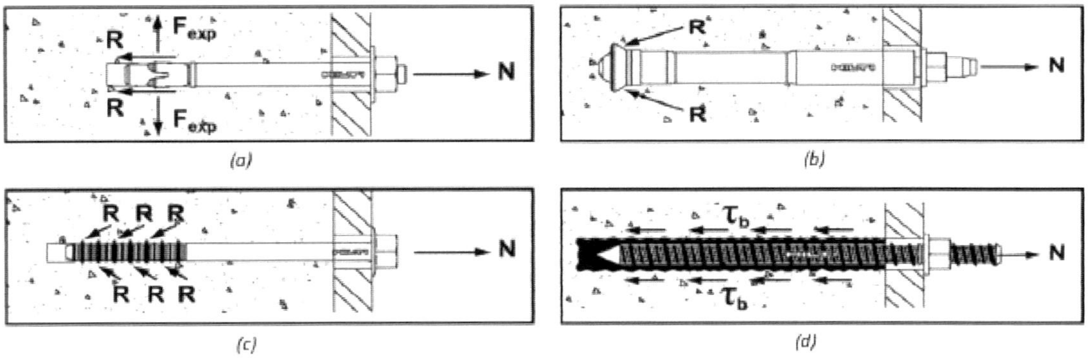

Figure 4. Types of bolts in the fixing systems: a) expanding bolts, b) undercutting bolts, c) screwed bolts, d) glued bolts [15].

The KOMAG experience [1, 4, 5] in the field of rock cutting tests with the use of mechanical bolts clearly shows that for this method, due to the nature of the loads application, the most reasonable is to use the undercutting bolts. The legitimacy of using undercutting bolts is due to the way the stress is exerted by the bolt, where all the stress caused by the applied force is concentrated at the place of the shape undercut at the bottom of the hole. Figure 5 shows an example of the undercutting bolt manufactured by HILTI.

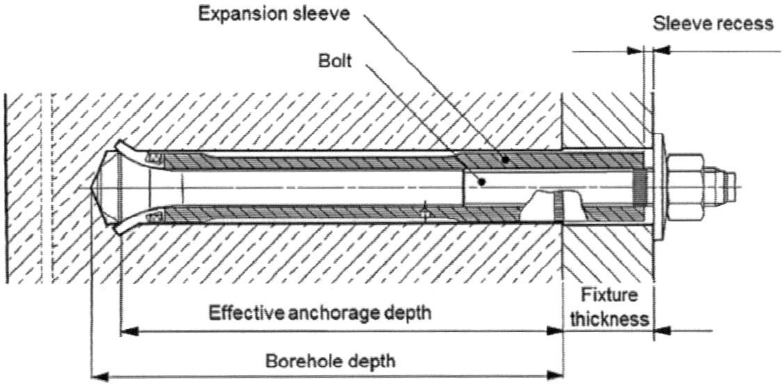

Figure 5. An example of the undercutting bolt [15]

Setting this type of bolt involves placing it in a hole previously prepared to the appropriate depth. Then, using a hammer drill and a special fixation device, torque and axial force from the impact acts on the expansion sleeve. Such load to the anchor sleeve causes it to expand on the conical end of the anchor while undercutting the bottom hole. After loading the bolt with an axial force directed to the hole bottom, the force begins to act on the walls of the undercut, which, after exceeding the critical value, creates a crack and causes its propagation. In Figure 6, the example of the loosening crack propagation with an illustration of the undercut residues at the beginning of its propagation.

IOP Conf. Series: Materials Science and Engineering **679** (2019) 012014 doi:10.1088/1757-899X/679/1/012014

Figure 6. Line of crack propagation obtained by undercutting bolt [13].

2.2. Test stand

The test stand for the rocks loosening in-situ tests consists of the device which has a simple and reliable structure, enabling manual transportation, assembly and operation. Construction of the testing device was adapted to each test conditions. The arrangement of the testing device for the loosening test is shown in Figure 7.

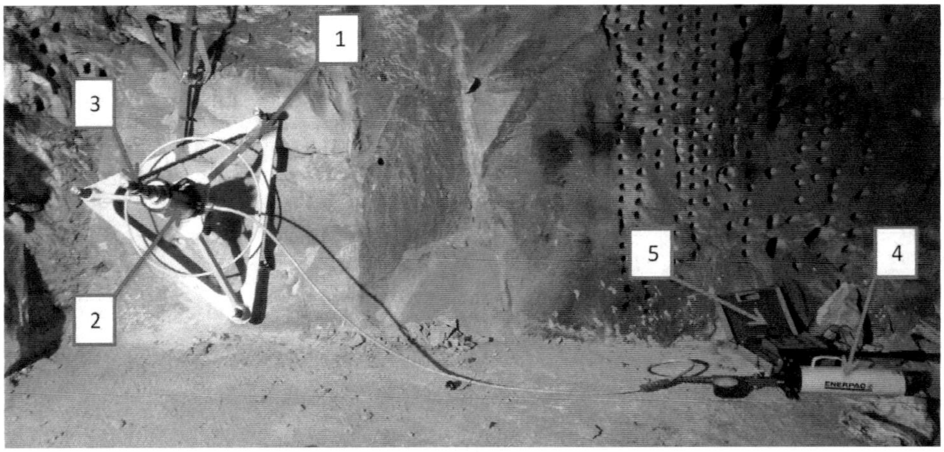

Figure 7. The RODEST test stand equipment.

As a standard, the testing device is equipped with the following components (Figure 7):

1. cylinder support,
2. cylinder with a cylindrical through hole,
3. undercutting bolts with fastening equipment,
4. supply unit (manual pump + hoses, connectors pressure gauge),
5. supply pressure recorder.

The cylinder support **1** enables installation of the cylinder **2** pulling out the fixed bolt. The support allows the use of two types of pull-out cylinders (30 and 60 T) by using a special connection plate. Each cylinder is supplied from the manual pump **4**. The support provides resting against solid rock at 3 points located on a diameter of 1000 mm. At each supporting point, there is the possibility of adjusting (adjustment in the range of 0÷180 mm) the supporting point to the unevenness of the operating front by tightening the pressure clamps. Pressure clamps also enable coaxial positioning of the cylinder **2** with a previously embedded bolt in the rock **3**. By using the pressure recorder **5** and known active surface of the cylinder **2**, it is possible to obtain the time process of changes in the pulling force during the rock loosening test. The test stand equipment is complemented by auxiliary equipment enabling efficient testing and recording the parameters, their changes and the results of loosening tests. Additional equipment components include: a hammer drill with accessories, a manual hoist and clamping ropes or a set of tools. The tests were recorded using a video camera, while the loosening measurements were taken using the hand-hold measuring instruments and a hand-held 3D scanner.

2.3. Place of tests

Tests of the loosening the rocks from a solid rock, due to the necessity of diversification of rock strength properties, were carried out in the following four mines, i.e. for four different types of rocks:

- The ZALAS open-cast porphyry mine. The tests were conducted on large rock blocks separated in the result of blasting the not damaged rock.
- The BRACISZÓW open-cast sand mine. The tests were carried out on large rock blocks separated as in the result of blasting.
- The GUIDO hard coal mine. The tests were carried out at the level of 320 m, at the longwall, on the mining front made in sandstone.
- The BRENNA open-cast sand mine. The tests was carried out on the mining front of the oblique stratified rock.

Each time, the rock material was collected from the testing site for the laboratory tests. In the laboratories of the Department of Geomechanics and Underground Construction of the Faculty of Mining, Safety Engineering and Industrial Automation of the Silesian University of Technology, strength tests of the material delivered by the KOMAG Institute of Mining Technology in Gliwice were carried out [16].

The mean values calculated from the test results are presented in Table 1:

Table 1. Results of laboratory tests of the material samples taken during in-situ tests [16].

Mean values of laboratory tests results	Uniaxial compression test, R_c, *[MPa]*	Uniaxial tensile strength, R_{rb}, *[MPa]*	Cohesion, c *[MPa]*	Angle of internal friction, φ *[°]*
Material / Mine				
Porphyry / ZALAS	106.5	5.9	8.6	54.0
Sandstone / BRACISZÓW	97.4	6.2	11.9	49.6
Sandstone / GUIDO	155.3	8.0	14.5	49.5
Sandstone / BRENNA	58.8	3.9	6.0	53.0

3. Results of in-situ tests

In-situ tests of the rock loosening process using an undercutting bolt was carried out in four different mines. As a result of the work carried out as part of the RODEST project, 175 tests were carried out to detach the rock material with a bolt fixed in the rock mass. The tests were carried out at various anchoring depths, thanks to which detachments of various volumes were obtained (Figure 8).

IOP Conf. Series: Materials Science and Engineering **679** (2019) 012014 doi:10.1088/1757-899X/679/1/012014

Figure 8. Sample fragments of loosened rock fragments for different anchoring depths [13].

Based on the data collected in the notes and video materials, the detailed procedure of each test was developed. From recording of the pressure in the cylinder, a graph of its changes and changes in force during the test cycle was developed. Geometric recreation of the loosening surface was made using a Shining3d hand-held 3D scanner. The process of manual scanning of the loosening surface is shown in Figure 9. The 3D cloud of points mapped in this way was used to develop virtual solid models of loosened rock cones. Data from the models made it possible to obtain information on the range and volume of loosened material. All collected data were placed in an individual test report chart and Excel spreadsheet.

Figure 9. Process of recreating the surface of the loosened material using the hand-held 3D scanner.

The surface of loosened material obtained using the scanned cloud of points allowed to generate for each test sample, cross-sections corresponding to the maximum (Z_{max}) and minimum (Z_{min}) range of the loosening propagation. An example of such a cross-section is shown in Figure 10. The generated cross-section also allowed for precise determination of the effective anchoring depth. The effective anchoring depth is determined by the depth at which the propagation of the loosening crack begins, what is shown as (H_{ef}) in Figure 10.

IOP Conf. Series: Materials Science and Engineering **679** (2019) 012014 doi:10.1088/1757-899X/679/1/012014

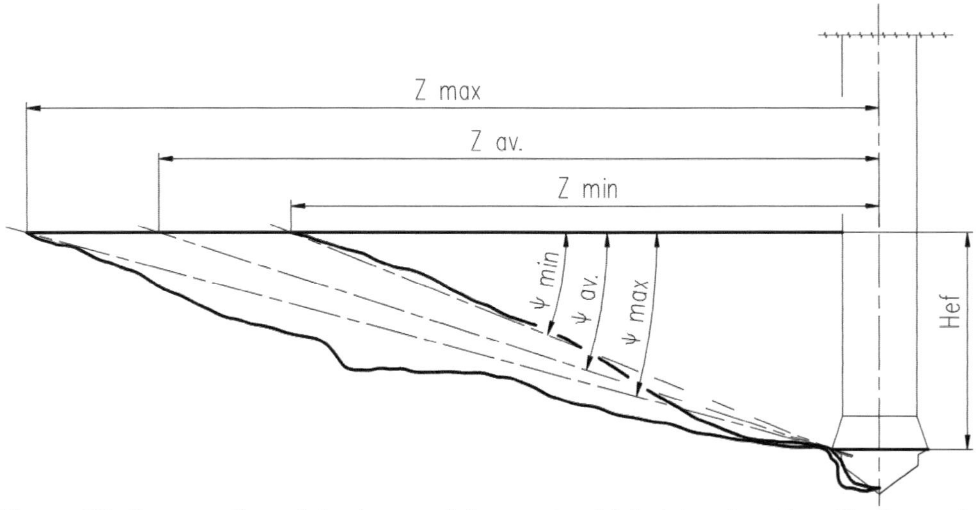

Figure 10. Cross-section of the loosened fragment, which determines the effective anchoring depth (H_{ef}), maximal (Z_{max}) and minimal (Z_{min}) loosening range.

The maximum force during each loosening test was determined by its highest value on the graph of force changes during the loosening test. A schematic determination of this value is shown in Figure 11. Such a graph was obtained by correlating the changes in cylinder pressure (measured with a connected recorder) to the effective cylinder surface area.

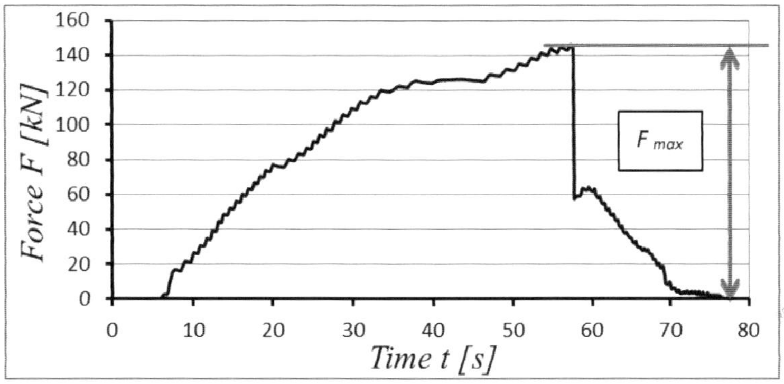

Figure 11. Determination of maximal loosening force (F_{max}) during the loosening test.

3.1. Analysis of impact of anchoring depth (H_{ef}) on the maximum loosening force (F_{max})

Assessment of the impact of effective anchoring depth on maximum force during loosening consisted in compilation of data collected during the tests, dividing them into the types of tested rocks. Based on the known depth of effective bolting and maximum force (F_{max}), all tests were put together. Presented results of each test were initially approximated by the exponential function divided into each mine (Figure 12). The figure also presents the approximation curve and the R^2 index.

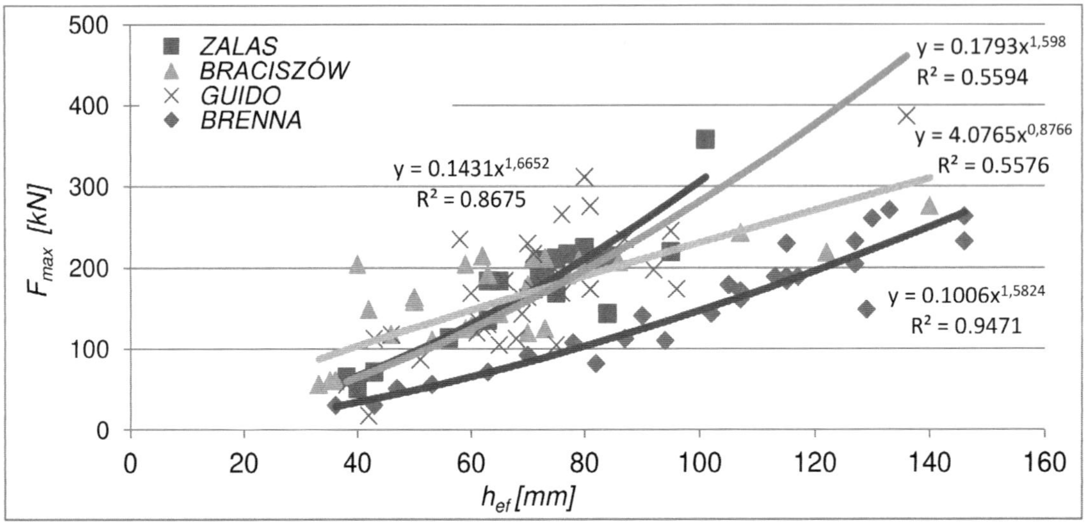

Figure 12. Maximum loosening force (F_{max}) in relation to the effective anchoring depth (h_{ef}) for different rock types

Analysing the approximating function for each rock type, it can be stated that for each rock, the loosening force increases with the anchoring depth. Exponential function with the exponent close to ~1.6 is the best type of function mapping. The exception is the function describing the test results in the BRACISZÓW mine, with exponent equal to ~0.9.

3.2. Determination of the average loosening range ($Z_{av.}$)

As it can be seen in Figure 10, for each scanned surface, the maximum (Z_{max}) and minimum (Z_{min}) range was determined, from which the average arithmetic value ($Z_{av.}$) was determined by the following equation:

$$Z_{av.} = (Z_{max} + Z_{min}) / 2 \quad (1)$$

Average loosening range ($Z_{av.}$) for each measurement depending on the effective anchoring depth (h_{ef}) is shown in Figure 13.

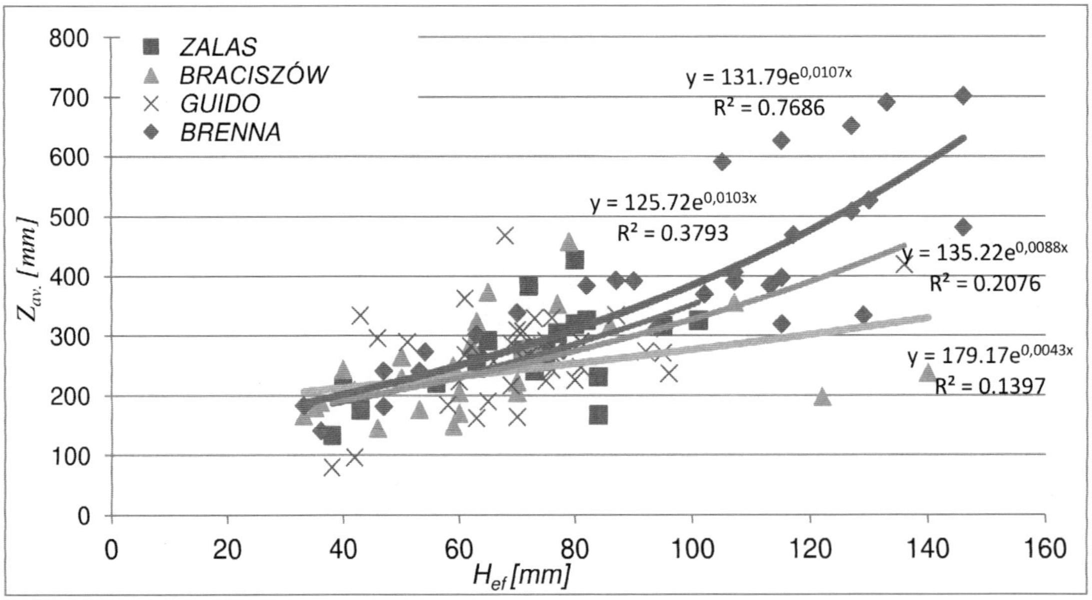

Figure 13. Average loosening range ($Z_{av.}$) for each measurement depending on the effective anchoring depth (H_{ef}) for different types of rocks.

The average loosening range for each type of rock was approximated by an exponential function. For each type of rock, an increase in the average loosening range can be seen, which is proportional to the increase in the effective anchoring depth *(H_{ef})*. For each test, the coefficient *(R)* determining the quotient of the average loosening range ($Z_{av.}$) and the effective anchoring depth (H_{ef}), determined by the following formula:

$$R = Z_{av.} / H_{ef} \tag{2}$$

In Figure 14 changes in coefficient *(R)* in relation to the effective anchoring depth for different types of rock are presented.

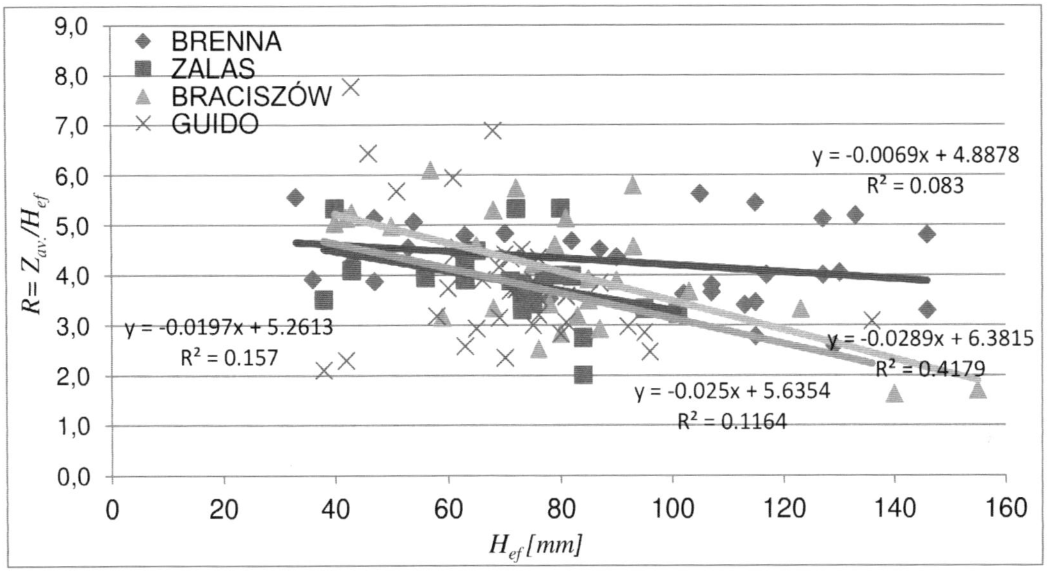

Figure 14. Changes in the coefficient *(R)* in relation to the effective anchoring depth *(H_{ef})* for different types of rock.

The average values of *(R)* coefficient for each type of rock within the anchoring depth range 30÷150mm do not differ much and are equal to 3.9 ÷ 4.2. These values are presented in Table 2.

Table 2. Average values of *R* coefficient.

	ZALAS	BRACISZÓW	GUIDO	BRENNA
$R = Z_{av.}/H_{ef}$	3.9	4.1	3.9	4.2

3.3. Loosening angle (ψ)

As it can be seen in Figure 10, the line connecting the beginning of the loosening crack propagation with its end was determined for each maximum *(Z_{max})* and minimum *(Z_{min})* range. For both lines, the loosening angle, referred to the base of the loosened cone i.e. maximum *(ψ_{max})* and minimum *(ψ_{max})* angles were determined. Then for both loosening angles, the average arithmetic value ($\psi_{av.}$) was calculated according to the following equation:

$$\psi_{av.} = (\psi_{max} + \psi_{min}) / 2 \tag{3}$$

Average values of loosening angles *($\psi_{av.}$)* for each test in relation to the effective anchoring depth *(H_{ef})* for different types of rock is shown in Figure 15.

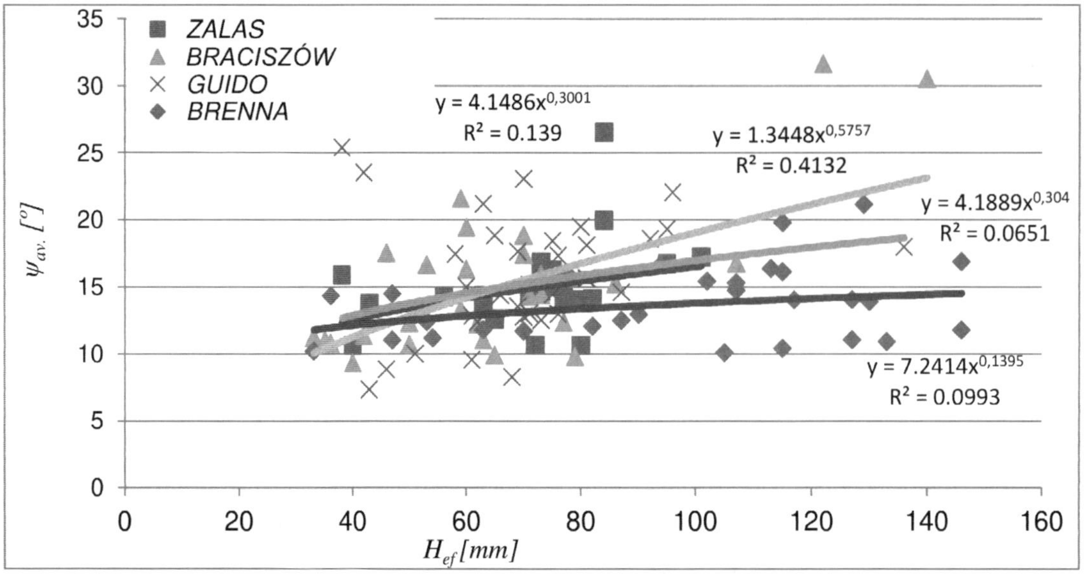

Figure 15. Average loosening angles ($\psi_{av.}$) for each test in relation to the effective anchoring depth (H_{ef}) for different types of rock.

Analysing the results of the average loosening angles, it can be stated that the average loosening angle increases ($\psi_{av.}$) with the effective anchoring depth (H_{ef}) for all tested rock types. The average values of the loosening angle for each rock type in the range of anchoring effective depth 30÷150 mm are presented in Table 3.

Table 3. Average values of the loosening angle $\psi_{av.}$.

	ZALAS	BRACISZÓW	GUIDO	BRENNA
$\Psi_{av.}[°]$	15.1	15.3	15.8	13.7

4. Conclusions

The tests, carried out so far within the RODEST project, indicate that the process of rock loosening using an undercutting bolt is a viable method for unconventional rock cutting. The current analysis of the results of the loosening process with the undercutting bolt allows the following general conclusions to be drawn:

- The maximum loosening force (F_{max}) needed to loosen a fragment of solid rock in a shape similar to a cone increases with the depth of effective anchoring (H_{ef}).
- The maximum force (F_{max}) depends on the strength properties of a given type of rock and increases with the increase of strength under uniaxial compression (R_c) and tensile strength (R_{rb}),
- Non-homogeneity or internal damages within the fixed bolt impacts the crack propagation and thus the loosening range in this direction. Therefore, the average values determined for the maximum and minimum loosening range from the axis of fixed bolt (Figure 10),
- The ratio of the average range value ($Z_{av.}$) and the effective anchoring depth (H_{ef}) determined by the (R) coefficient is approximately 4. This is more than twice the value used for the calculation of the anchoring capacity (Figure 2). At the same time, one can notice a tendency to reduce the (R) coefficient together with increase of the effective anchoring depth for all types of tested rocks.
- Average values of loosening angle ($\psi_{av.}$) for each rock type, within the range of tested anchoring depths varies from 13-17°. For each type of rock, there is a tendency to increase the loosening angle ($\psi_{av.}$) with the increase of effective anchoring depth (H_{ef}).

The conducted tests as well as analysis of the results collected during the tests allow to undertake a research work aimed at building a simplified loosening model that allows estimation of loosening forces and range, depending on the anchoring depth and rock strength in order to enable correct realization of the rock loosening process in mine conditions. The in-situ tests allow the adaptation of the numerical simulation method best suited to the real tests.

Acknowledgments
The article has been developed within the RODEST project, OPUS 10 competition No. 2015/19/B/ST10/02817, financed by the National Science Centre (Project No. 2015/19/B/ST10/02817).

References
[1] Siegmund M and Jonak J Analiza wyników badań wstępnych oraz określenie kierunku dalszych prac badawczych technologii drążenia tuneli ratowniczych metodą mechanicznego odspojenia. *Cuprum* 2017 no **1** s. 57-71
[2] Klich A et al. *Niekonwencjonalne techniki urabiania skał* Wydawnictwo Śląsk, Katowice1998
[3] Kalita M and Prostański D Technologia drążenia tuneli ratowniczych metodą niszczenia spójności skał *Przegląd Górniczy* 2012 no **12** s. 86-91
[4] Cebula D Wyniki badań dołowych technologii mechanicznego odspajania skał. Nowoczesne metody eksploatacji węgla i skał zwięzłych. Monografia, Akademia Górniczo-Hutnicza im. Stanisława Staszica, Kraków 2015 s. 212-222
[5] Drwięga A et al. Sprawozdanie za rok 2014 z realizacji projektu INREQ ITG KOMAG. Gliwice 2014 (not published)
[6] Ashour A and Alqedra M Concrete breakout strength of single anchors in tension using neural networks. *Adv. Eng. Softw.*, **36** (2) (2005), pp. 87-97
[7] Fuchs W, Eligehausen R and Breen J E Concrete capacity design (CCD) approach for fastening to concrete, *ACI Struct J*, **92** (1) (1995), pp. 73-94
[8] ACI Committee 355. State-of-the-art report on anchorage to concrete. Detroit, USA: American Concrete Institute; 1991. Reapproved 1997
[9] Comite Euro-International du Beton (CEB), Task Group VI/5. Fastenings to concrete and masonry structures. London, UK: ThomasTelford Services Ltd; 1994. p. 249.
[10] Tan E L, Varsani H and Liao F -Y (2019). Experimental study on demountable steel-concrete connectors subjected to combined shear and tension. *Engineering Structures*, **183**, 110-123. https://doi.org/10.1016/j.engstruct. 2018.12.08
[11] Jiabao Y (2012-05-30) Ultimate Strength Behaviour of Steel-Concrete-Steel Sandwich Composite Beams and Shells.. ScholarBank@NUS Repository.
[12] Brincker R, Ulfkjær J P, Adamsen P, Langvad L and Toft R (1995) Analytical Model for Hook Anchor Pull-out. In R. Widmann (Ed.), Proceedings of the International Symposium on Anchors in Theory and Practice: Salzburg, Austria, 9-10 Ooctober 1995 (pp. 3-15), CRC Press/Balkema
[13] Jonak J, Siegmund M et al. Sprawozdanie z realizacji zadnia 2 projektu RODEST, ITG KOMAG. Gliwice 2018 (not published)
[14] Gontarz J, Podgórski J and, Siegmund M Comparison of crack propagation analyses in a pull-out test, CMM 2017, 22nd International Conference on Computer Methods in Mechanics, Lublin 13-16 September 2017
[15] Anchor Fastening - Technology Manual, Hilti Corporation, 09 / 2014
[16] Tomiczek K Badania wytrzymałościowe wybranych skał magmowych i osadowych. Raporty z badań dla ITG KOMAG: NB-363/RG4/2017, NB-11/RG4/2017, NB-117/RG4/2018, NB 266/RG-4/2018, Katedra Geomechaniki i Budownictwa Podziemnego, Wydział Górnictwa, Inżynierii Bezpieczeństwa i Automatyki Przemysłowej, Politechnika Śląska, Gliwice 2017-2018

IOP Conf. Series: Materials Science and Engineering **679** (2019) 012015 doi:10.1088/1757-899X/679/1/012015

The method of determining the optimal life cycle of the mining machine

A Kowalski[1]* **and M Blichtarski**[1]

[1] Wroclaw University of Science and Technology, Faculty of Mechanical Engineering, 27 Wybrzeże Wyspiańskiego St., 50-370 Wrocław, Poland

*E-mail: arkadiusz.kowalski@pwr.edu.pl

Abstract. One of the significant components of production costs are the purchase and operating costs of machinery and equipment; according to Sandvik Coromant, they can account for approx. 30% of manufacturing costs. In the mining industry, the share of purchase and operation costs of machinery and equipment in the mining cost is certainly even greater. Such a large share of machine and tooling costs in manufacturing costs determines the significance of the issue of minimizing these costs, which is connected with the problem of determining the moment when further operation of a worn machine becomes unprofitable. One of the currently available ways to determine the length of the operation lifecycle is to register and compare the total costs of machine maintenance and its annual maintenance costs. This method is easy to apply, unfortunately it is not possible to use it at the machine purchase planning stage and thus it does not allow for prior determination of a viable life cycle length. The aim of the article is to propose a method for determining a profitable life cycle of a mining machine, taking into account the economic factors of macroenvironment. The costs related to operation, including failure costs, were also considered. Determining the duration of the operation cycle is based on determining the payback time of investment for the purchase of a new machine or on minimizing the average cost of monthly use and purchase of the machine in several operating cycles for a longer period of operation. The differences in operating costs between used and new machines are taken into account in calculating the payback period. Sample calculations were carried out using the example of the self-propelled Getman S330 scaler.

1. Introduction

One of the criteria that can be used to estimate the profitability of an investment may be its payback period. This parameter is widely used to evaluate investments [1, 2]. It defines the time which is necessary for the costs incurred for the investment to be paid back. It plays a special role when financial liquidity is the key factor for the company [3]. The calculated investment payback period parameter can be a key factor in choosing between two investments. An investment with a shorter payback is considered a less risky investment. If there is no need to choose between two projects, the payback period should be referred to the reference payback period set by the company [4].

The payback period is calculated as:

$$PBP = \frac{\text{The cost of investment}}{\text{Annual income from investment}} \quad (1)$$

where: PBP – the payback period [5].

The use of the payback period parameter according to formula 1 has the following advantages:

- ease of using the payback period parameter,
- it enables comparison of projects in terms of liquidity risk,
- low-cost method calculations affect low analysis costs.

In the case of a new machine purchase, the proceeds from the investment will not have a fixed value, therefore it is impossible to apply the formula 1 in its current form. The proceeds from the investment will be determined as the difference in the costs incurred in relation to the maintenance of the machines. The payback period will be determined as the moment when profits from the purchase of a new machine exceed its cost.

1.1. Life Cycle Cost Analysis

By definition, the life cycle cost analysis (LCC - Life Cycle Cost) is a process aimed at determining the sum of costs related to a product or a project, including costs of acquisition, installation, operation, repairs and utilization. It is a tool for estimating the total costs that appear during the life cycle. The LCC method was developed by the Department of Defence of the United States in the mid-1960s and was used in large infrastructure projects like military centres and refineries. In later years, different variations of the method were developed, such as the Environmental Life Cycle Cost (ELCC). The main purpose of the LCC method is to compare the total life cycle costs for different products or alternative processes. It is a method to identify the most cost-effective variant in order to facilitate the decision-making process.

In the life cycle cost analysis method, you can distinguish the following four steps:

- cost structure analysis – this step is to identify all elements affecting costs in the life cycle and determine the limits of the application of the analysis. This is the most important step because its correct implementation prevents overlooking or double-counting of specific costs,
- cost estimation – the process of determining the approximate value of costs based on known values, values predicted based on empirical data or based on expert opinions,
- discounting – this is the process of calculating the value of a given amount of money based on its value in the future. If the value of the discount rate is high, it means that the money currently held is more important, if it is lower, more important is the money held in the future. If the discount rate is zero, it means that there is no difference between the value of same amount of money at different times. The value of the discount rate depends on many factors, such as inflation, investment opportunities, preferences and should be determined when consulting an expert.
- inclusion of inflation – inclusion of price changes in the considerations [6].

However, it is difficult to decide whether to end the use of the machine or the device based on the payback period parameter or to seek an answer to this question directly in the LCC method.

2. Main factors affecting the economic aspects of machine operation

The most important factors occurring regardless of the type of machine used were selected. They include:

- the purchase cost of a new machine or a device,
- operating costs,
- repair costs,
- influence of inflation.

These factors will have a decisive impact on the method proposed in the following chapters to determine the optimal life cycle of the machine.

2.1. Operating and repair costs

One of the types of costs that change during the operation of the machine are the costs of repairs and operation [7]. Many factors affect such costs, including the environment of exploitation, care for the machine and its equipment. Two groups of processes that influence the evolution of the machine state can be distinguished:

- physical aging,
- machine wear.

Physical aging is a term used for physical processes that lead to irreversible changes of properties of the operated object. They occur as a result of both external and internal extortion, starting from machine production and leading to gradual loss of physical properties.

Machine wear means permanent, undesirable changes occurring during its operation. They are related to the forces occurring in interacting kinematic pairs during the transformation of energy into mechanical work. Variable mechanical stresses created during operation are responsible for the wear processes. There are three phases that can be distinguished in the course of wear (Figure 1):

- Phase I – called also the break-in period. It is a short-term period, and its duration is not clearly definable. During it, a significant intensity of wear caused by material defects and production errors is observed. Wear in this phase is not harmful and leads to the transformation of the state of layers of cooperating elements to a state characterized by higher wear resistance;
- Phase II – it is characterized by a constant intensity of wear. This phase is responsible for the period of proper cooperation of elements;
- Phase III – the period of accelerated wear. This is the period during which the intensity of wear increases. This is related to the increased clearance between the cooperating elements and the associated occurrence of additional dynamic loads. They result in larger deformations of elements leading to change of tribological conditions and increase of material losses [8, 9].

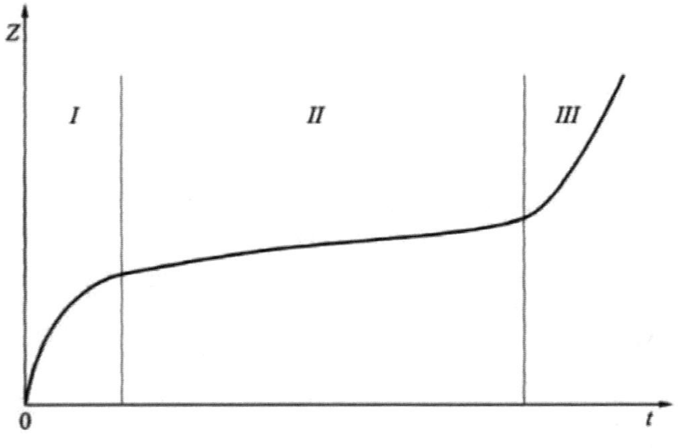

Figure 1. Empirical wear curve according to Lorenz, where: Z – wear, t – time [8].

The mathematical model that can be used in estimating the costs of repairs and operation can be formula 2 used in the studies of Markus Lips and Frank Burose [10]:

$$y = \beta_0 \prod_{i=1}^{n} x_i^{\beta_i} \tag{2}$$

where: y – annual cost of repairs and operation expressed as a part of the machine price, x_i – factors affecting the costs of repairs and operation, β_i – coefficients.

The annual use of the machine and its age are the factors that most strongly affect the costs of repairs and operation. At the same time, they were the only universal factors (used for each type of the machines considered in the study). There is a relationship between the degree of use of the machine and the total cost of repairs and operation, assuming a constant total working time. In the case of longer service cycles, the total costs of repairs and operation may be in extreme cases 3 ÷ 4 times higher than in short ones. This indicates that the costs of repairs and operation will be the smallest with short operating cycles [11]. Similar conclusions can be drawn by analysing data on the costs of repair and maintenance of agricultural tractors. They show the relationship between the costs of using the machine and the age of the tractor. On this basis, it can be stated that these costs at a certain point in the operating cycle may be 4 ÷ 5 times higher than at the beginning [12].

There is also a group of methods – indirectly helpful in deciding on the end of machine operation – for calculations of profitability of overhaul (e.g. balance account [13], maximum limit cost method [14]) and optimization of machine replacement period [15], developed in 70's and 80's of the 20th century. They enable determining the economic limit of outlays for overhaul, based on the analysis of costs incurred during the actual operation of the machine. However, it is impossible or difficult to use these methods at the stage of planning the machine purchase.

2.2. Inflation

In the case of long-term considerations, it is necessary to take into account the prices that are changing on the market. Not only the costs of machines, but also the costs of consumables change over time, which results in an increase in operating costs.

One of the price change indicators commonly used in the world is PPI - Producer Price Index. This is the oldest of this type of indexes. It was introduced in the United States, where its calculation and publication is dealt with by the Bureau of Labour Statistics as part of the US Department of Labour. In Poland, PPI takes the form of the price index of production sold in the industry, and its calculation and publication of results is handled by the Central Statistical Office. The price index of production goods for the European Union, the eurozone and individual Community countries is published by Eurostat.

The PPI index is usually calculated as the ratio of prices in a given year to prices in the base year. The US Labour Statistics Office in the calculation of PPI covers all branches of industry excluding imports.

3. Description of the machine subject to analysis – Getman S330

The subject of the analysis is the life cycle of the Getman S330 mining machine (Figure 2); this device is used to secure excavations by removing loose rocks hanging from the ceiling or side walls. Such a treatment is necessary to ensure the safety of people as well as to protect the equipment from damage caused by the detachment and fall of rocks, which could cause not only an accident, damage to property, but also halt the mining [16].

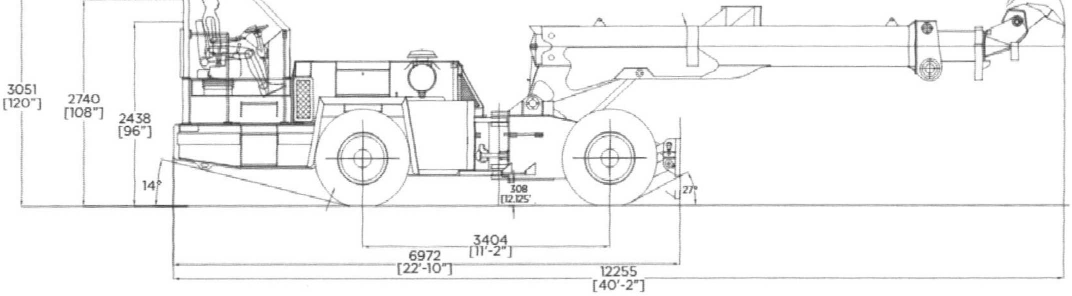

Figure 2. Mining machine Getman S330, main dimensions in mm [16].

The Getman S330 machine is equipped with a 4-cylinder engine with 173 horsepower, and its reach enables works at a height of up to 9 meters. The estimated cost of the machine is PLN 2 900 thousand (€ 680 thousand). In the considered case of the mining plant, the device was set up in a 2-shift system for 250 days a year. It was assumed, based on historical data, that the average annual use of the machine is 56%. This data is necessary to calculate the hourly use of the machine during the year.

4. Algorithm of the method for determining the optimal time of machine operation cycle

Figure 3 shows the algorithm of the proposed method for determining the optimal cycle time of the machine. The method has the following steps:

1. selection of coefficients and calculation of the machine usage rate,
2. construction of a table of repair costs, repair costs including inflation, accumulated repair and operating costs, and machine prices including inflation.

The next steps of the method depend on the adopted optimization criterion. The method assumes a choice between minimizing the average costs of operating cycles per unit of operating time and optimization based on the payback period. The payback period criterion should be chosen if the priority is to ensure financial liquidity. In situations where the investment does not have a significant impact on the financial liquidity of the enterprise, the criterion of average costs of operating cycles should be selected.

For the criterion of average costs of operating cycles, it is necessary to take the following steps:

3a. building a table of machine purchase moments,
4a. building a table of machine purchase costs,
5a. building a table of cumulative repair and operating costs,
6a. calculation of the sum of costs and average costs of operating cycles per unit of operating time, finding the minimum.

For the criterion of the payback period, the subsequent steps take the following form:

3b. defining the target payback period based on investment plans and the company's investment strategy,
4b. calculation of monthly profits resulting from lower costs of repair and operation of a new machine,
5b. calculation of the sum of profits since the purchase of a new machine,
6b. finding the payback period for each moment of purchase,
7b. finding the optimal purchase moment based on the target payback period.

Step 1 – selection of coefficients and calculation of the machine usage rate

The first step of the method is the selection of coefficients and the calculation of the hourly use of the machine. Coefficients should be selected depending on the machine type according to the literature data, manufacturer's recommendations or actual data on the operating costs of the machine already in use. The hourly usage of the machine can be calculated according to the formula 3:

$$W_h = l_z l_d W_{\%} \tag{3}$$

where: W_h – hourly use of the machine during the year [h], l_z – number of shifts, l_d – number of working days per year, $W_{\%}$ – machine usage [%].

At this stage, it is also necessary to specify the time of operation under consideration (it was assumed that $t_e = 600$ [month]) and inflation indicators.

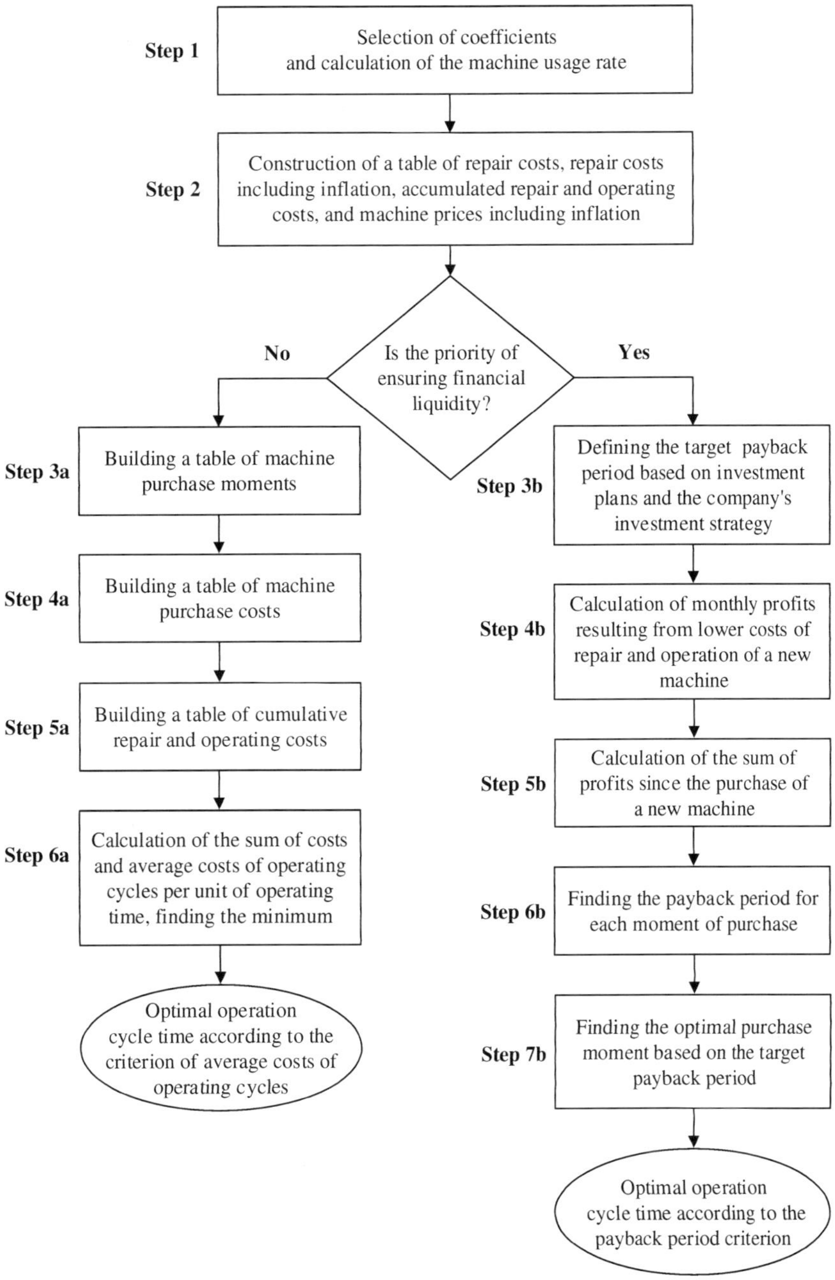

Figure 3. The algorithm of the method for determining the optimal time of the machine's operating cycle.

Step 2 – construction of a table of repair costs, repair costs including inflation, accumulated repair and operating costs, and machine prices including inflation

In order to calculate the costs of repairs, repair costs including inflation, cumulative repair and operating costs, and machine prices including inflation, formulas 4 to 7 are needed. Formula 4 is used to determine the costs of repairs and exploitation in the *i*-th month of machine operation:

$$K_{nie_i} = \frac{c_m \beta_0 \left(\frac{i}{12}\right)^{\beta_1} W_h^{\beta_2}}{12} \tag{4}$$

where: i – month of operation, K_{nie_i} – costs of repairs and operation in a given month [PLN], C_m – machine cost [PLN], β_0, β_1, β_2 – coefficients influencing the costs of repairs and operation, W_h – [h hourly use of machines during the year [h].

Formula 5 is used to calculate the costs of repairs and operation including inflation. These costs are calculated on the basis of the formula for compound interest:

$$K_{nie(I)_i} = K_{nie_i}(1 + I_{nie})^{\frac{i}{12}} \tag{5}$$

where: $K_{nie(I)_i}$ – costs of repairs and operation including inflation in the i-th month [PLN], I_{nie} – inflation coefficient for repair and operating costs.

Formula 6 enables calculation of cumulative repair and operating costs. The cumulative cost of repairs and operation is the sum of the costs of repairs and operation, taking into account inflation, from the beginning of the machine's use until the i-th month:

$$K_{nie(\mathrm{skum})_i} = \sum_{k=1}^{i} K_{nie(I)_k} \tag{6}$$

where: $K_{nie(\mathrm{skum})_i}$ – the cumulative cost of repairs and operation for the i-th month [PLN].

Formula 7 allows you to calculate the cost of purchasing a new machine in the i-th month, including inflation:

$$C_{m(I)_i} = (C_m - W_m(i))(1 + I_m)^{\frac{i}{12}} \tag{7}$$

where: $C_{m(I)_i}$ – the cost of purchasing a new machine, regarding inflation [PLN], I_m – inflation coefficient for the price of the machine, $W_m(i)$ – the value of the worn-out machine as a function of its age [PLN].

4.1. The algorithm of the method for the criterion of average lifecycle costs

Step 3a – building a table of machine purchase moments

To build the table of the machine purchase moments, it is necessary to use formulas 8 and 9. Formula 8 is used to determine the end of machine use. This is the month of withdrawal from use of a given type of machine, assuming a full number of operating cycles of the length of i months, hence the result is rounded up:

$$T_{e_i} = \left\lceil \frac{t_e}{i} \right\rceil i \tag{8}$$

where: T_{e_i} – end of machine operation, t_e – assumed last month of machine operation, i – length of the operating cycle.

Formula 9 is used to determine the purchase moments of a new machine. For every length of the operation cycle, the first purchase of the machine will be made at the very beginning of the cycle, i.e. in the zero month. The next purchase of a new machine will take place in a month being a multiple of the life cycle length, shorter than the last month of machine operation:

$$M_{z_{i,j}} = i(j-1), j: j \in N \wedge j \leq \frac{T_{e_i}}{i}, i: i \in N \wedge i \leq t_e \tag{9}$$

where: $M_{z_{i,j}}$ – j-th moment of purchase of a new machine for the length of the operation cycle i.

Step 4a – building a table of machine purchase costs

The construction of the machine purchase cost table was based on the formula 10. This expression was used to determine the purchase costs of the machine in the j-th operating cycle for the cycle length i. This cost is equal to the purchase cost of the machine at the corresponding purchase moment $M_{z_{i,j}}$:

$$C_{m(I)_{i,j}} = C_{m(I)_{M_{z_{i,j}}}}. \tag{10}$$

Step 5a – building a table of cumulative repair and operating costs

To build a table of cumulative repair and operating costs, it is necessary to use formulas 11 and 12. Formula 11 is used to determine the cumulated costs of repairs and operation of the machine in the j-th operation cycle for the cycle length i. This cost is equal to the cumulative cost of repairs and operation of the machine with length cycle i and taking into account inflation:

$$K_{nie(\mathrm{skum})_{i,j}} = K_{nie(\mathrm{skum})_i}(1 + I_{nie})^{\frac{M_{z_{i,j}}}{12}} \tag{11}$$

where: $K_{nie(\mathrm{skum})_{i,j}}$ – the cumulative cost of repairs and operation of the machine from the j-th operating cycle of length i [PLN].

Formula 12 makes it possible to determine the total repairs and operation costs incurred for j operating cycles of length i. This cost is equal to the sum of accumulated costs of repairs and operation of machines with cycle length i:

$$K_{niec(\mathrm{skum})_i} = \sum K_{nie(\mathrm{skum})_{i,j}} \tag{12}$$

where: $K_{niec(\mathrm{skum})_i}$ – total repair and operating costs for the cycle length i [PLN].

Step 6a – calculation of the sum of costs and average costs of operating cycles per unit of operating time, finding the minimum

The formulas 13 and 14 are needed to complete the table of total costs and average costs per unit of operation time. Formula 13 defines the total costs incurred for the operation cycle of the length i. This cost is equal to the sum of the costs of repairs and operation and the purchase costs of machines:

$$K_{c_i} = K_{niec(\mathrm{skum})_i} + C_{mc(I)_i} \tag{13}$$

where: K_{c_i} – total costs for the exploitation cycle of length i [PLN].

Formula 14 is used to determine the average costs of operating cycles per unit of operation time:

$$K_{cś_i} = \frac{K_{c_i}}{T_{e_i}} \tag{14}$$

where: $K_{cś_i}$ – average costs of operating cycles per unit of operation time [PLN].

After calculating the average costs of operating cycles per unit of operating time, it is possible to determine the optimum cycle time $t_{c_{opt}}$. This is the time of cycle i for which the average costs of operating cycles will be the smallest (Formula 15):

$$t_{c_{opt}} = \{i: K_{cś_i} = \min\{K_{cś}\}\}. \tag{15}$$

4.2. The algorithm of the method for the payback period criterion

Step 3b – defining the target payback period based on investment plans and the company's investment strategy

As a result of this step, the optimal payback period PBP_{opt} is determined.

Step 4b – calculation of monthly profits resulting from lower costs of repair and operation of a new machine.

Formula 16 is used to determine the profits resulting from the lower costs of repair and operation of a new machine in the i-th month of its use. Profit is the difference between the costs of repairs and exploitation of a worn out machine and the repair and operation costs of a new machine:

$$Z_{i,j} = K_{nie_{i+j}} - K_{nie_i} \tag{16}$$

where: $Z_{i,j}$ – profit resulting from lower costs of repairs and operation in the i-th month of use of the machine purchased after the j-th month of operation of the machine that has been depleted [PLN].

Step 5b – calculation of the sum of profits since the purchase of a new machine

Formula 17 made it possible to calculate the total profits resulting from the lower repair and operating costs of a new machine in the *i*-th month of its use. This is the sum of profits until the *i*-th month of use of the machine.

$$Z_{i,j} = K_{nie_{i+j}} - K_{nie_i} \tag{17}$$

where: $Z_{c_{i,j}}$ –total profits resulting from lower repair and operating costs [PLN].

Step 6b – finding the payback period for each moment of purchase

Formula 18 allows you to set a payback period when purchasing a new machine after *j* months of using a worn out machine. The payback period is the minimum time in a month, after which the profits will be at least equal to the costs of purchasing a new machine, reduced by the value of a worn-out machine.

$$PBP_j = \min\{i: Z_{c_{i,j}} \geq C_m - W(j)\} \tag{18}$$

where: PBP_j – the payback period for the purchase of a new machine after *j* months of use of a worn out machine.

Step 7b – finding the optimal purchase moment based on the target payback period

The optimal operation cycle time determined on the basis of the payback period is equal to the smallest month of purchase of a new machine, for which the payback period is equal to the optimal one:

$$t_{c_{opt}} = min\{j: PBP_j = PBP_{opt}\} \tag{19}$$

where: $t_{c_{opt}}$ – optimal operation cycle time.

5. Presentation of calculation results on the example of Getman S330

In order to apply the algorithm for sample calculations, the above formulas were included in the MS Excel spreadsheet. Missing coefficients were selected based on literature data; they were collected in Table 1.

Table 1. Selected values of coefficients.

Name	Symbol	Unit	Value
The price of the machine	C_m	thousands PLN	2 900
The value of the worn-out machine	$W_m(i)$	thousands PLN	580
Assumed operating time	t_e	month	600
Coefficient β_0	β_0	%	5
Coefficient β_1	β_1	–	1.4
Coefficient β_2	β_2	–	0.07
Number of shifts	l_z	–	2
The number of days of the machine usage	l_d	–	250
Machine usage rate	$W_{\%}$	%	56
Inflation rate for the price of the machine	I_m	%	3.5
Inflation rate for repair and operating costs	I_{nie}	%	2.35
Optimum payback period	PBP_{opt}	month	12

5.1. Calculation results for the criterion of average costs of operating cycles.

The final results of the algorithm are presented in Figures 4 and 5. These charts show the average costs of operating cycles per unit of operating time in connection with the purchase of machines and their repair and operating costs for the total number of operating cycles with length *i* in operating time greater than or equal to the assumed. There are noticeable large disparities between average costs, especially in the case of very short and very long operating cycles (Figure 4). The smallest costs occur in the range of operating cycles up to 100 months.

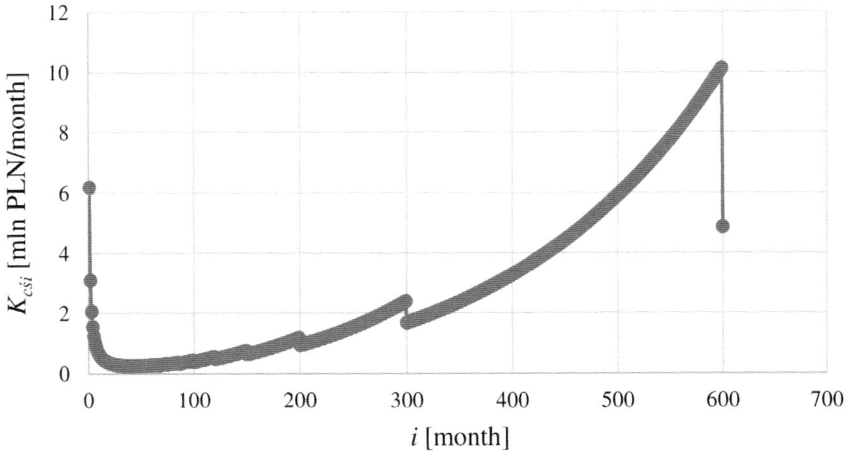

Figure 4. The chart of average costs of operating cycles.

Due to the insufficiently accurate scale to read the minimum, an additional chart was prepared (Figure 5), presenting a section of data near the minimum of the global chart. The chart also shows the total costs of machinery purchase and the total costs of repairs and operation.

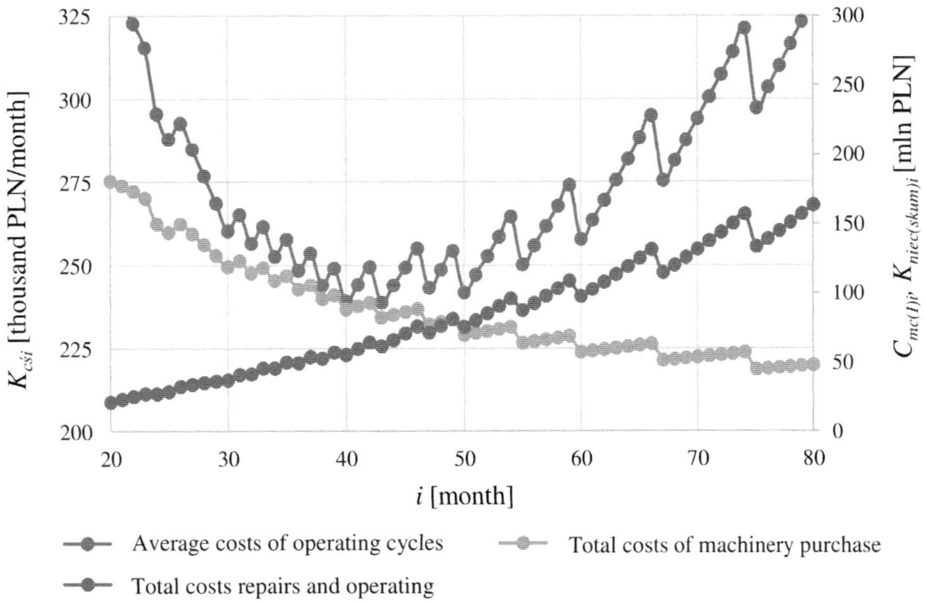

Figure 5. The chart of average costs of operating cycles, total costs of machinery purchase and total repairs and operating costs.

The minimum readout has been achieved for the cycle length i = 43 months and is equal 239 thousand PLN/month. The smallest values of average costs of operating cycles are in the range in which the total purchase costs of machines and the total costs of repairs and operation reach similar values. Significantly, the global minimum of average life-cycle costs has not been achieved for the intersection of operating costs curves and the purchase of new machines. Therefore, for the criterion of average costs of operating cycles, $t_{c_{opt}} = 43$ months is the optimum cycle time.

5.2. Calculation results for the payback period criterion

The vector of payback period PBP_j is presented in Figures 6 and 7. The payback period reaches the highest values for the shortest operating times *i* (for 100 ÷ 400 months).

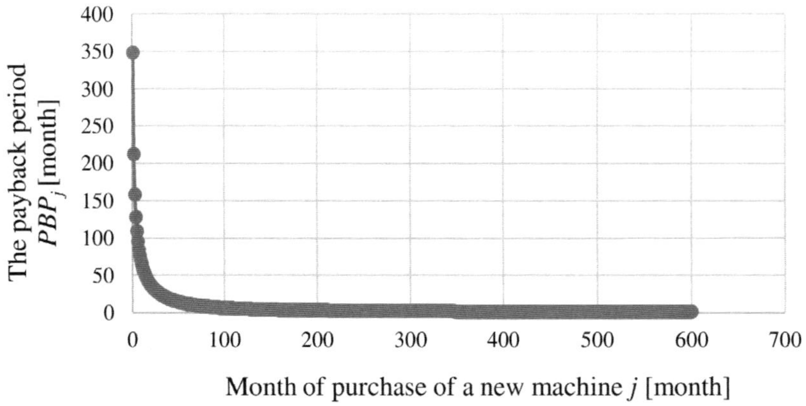

Figure 6. The payback period PBP_j.

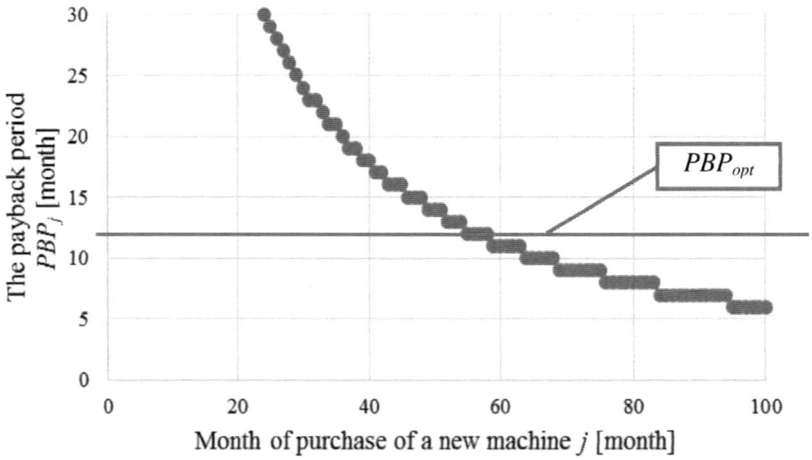

Figure 7. The payback period PBP_j, intersection point.

The values of the PBP_j vector are arranged in non-incremental order. The red line in Figure 7 indicates the assumed optimal payback period. The optimal operation cycle time $t_{c_{opt}}$ determined for this criterion is 55 months.

6. Summary

Methods for computational optimization of the machine life cycle length, developed mostly in the 60s, 70s and 80s of the twentieth century, are difficult to access or are not popularized. These methods, used to determine the profitability limit of overhauls, focus on the analysis of costs incurred during the actual operation of the machine. This approach makes it impossible to use these methods when planning a machine purchase.

A significant share of machine and tooling costs in manufacturing costs indicates the importance of this issue. The selection of the length of time of the machine operation cycle at the planning stage can have a significant impact on the profitability assessment and making investment decisions. The proposed method has the following advantages:

- the possibility of using the method already at the stage of planning the purchase of the machine,
- takes into account the costs of using the machine at different lengths of operating cycles,
- inclusion of purchase, operation and repair costs as well as economic factors of macroenvironment.

The general nature of the method means that it can be applied to most machines and devices. The presented formulas can be extended with further factors in special cases affecting the operating costs.

References

[1] Gola A 2014 Economic aspects of manufacturing systems design *Actual problems of economics* 205–12
[2] Gola A 2018 Reliability analysis of reconfigurable manufacturing system structures using computer simulation methods *Eksploatacja i Niezawodnosc - Maintenance and Reliability* **21** 90–102
[3] Nowakowski G A and Hahn M P 2013 Forecasting the Future of Alternative Energy Technologies Using Economic Payback Curves *Distributed Generation & Alternative Energy Journal* **28** 28–52
[4] Pinkasovitch A An Introduction to Capital Budgeting *Investopedia*
[5] Kwat N 2016 Evaluating Profitability of Investment Projects *Economics Discussion*
[6] Kara S 2014 Life Cycle Cost CIRP *Encyclopedia of Production Engineering* (Berlin, Heidelberg: Springer Berlin Heidelberg) pp 751–6
[7] Jasiulewicz-Kaczmarek M and Żywica P 2018 The concept of maintenance sustainability performance assessment by integrating balanced scorecard with non-additive fuzzy integral *Eksploatacja i Niezawodnosc - Maintenance and Reliability* **20** 650–61
[8] Kalandyk B 2002 Zastosowanie matematycznego modelu kinetyki zużycia skumulowanego w warunkach testu Millera *Archives of Foundry* **2** 376–83
[9] Tylicki H and Żółtowski B 2012 *Genezowanie stanu maszyn* (Radom: Wydawnictwo Naukowe Instytutu Technologii Eksploatacji - PIB)
[10] Beppler D and Hummedia M 1985 Repair cost of agricultural machinery in developing countries *American Society of Agricultural Engineers*
[11] Lips M and Burose F 2012 Repair and Maintenance Costs for Agricultural Machines *International Journal of Agricultural Management* **1** 8
[12] Abdelmotaleb I A 1989 *Repair and maintenance cost analysis of tractors and combines* (Iowa State University)
[13] Zbichorski Z 1971 *Opłacaność remontów maszyn i urządzeń* (Warszawa: PWE)
[14] Wilczkowski D and Gontarski D 2011 Planowanie wymiany lub remontu wybranego obiektu technicznego z wykorzystaniem metody maksymalnej kwoty *Zeszyty Naukowe Politechniki Świętokrzyskiej. Nauki Techniczne* **Z. 16** 121–6
[15] Piasecki S 1972 *Optymalizacja systemów obsługi technicznej* (Warszawa: Wydawnictwa Naukowo-Techniczne)
[16] Getman Product Datasheet 2015 Mechanical Scaler S330

IOP Conf. Series: Materials Science and Engineering **679** (2019) 012016 doi:10.1088/1757-899X/679/1/012016

The use of innovative solutions improving selected energy or environmental indices of hydrostatic drives

Z Domagała[1]*, K Kędzia[1] and M Stosiak[1]

[1] Wrocław University of Science and Technology, Faculty of Mechanical Engineering, Ignacego Łukasiewicza 5, 50-371 Wrocław, Poland

*E-mail: zygmunt.domagala@pwr.edu.pl

Abstract. This paper consists of three parts. The first part presents an exemplary application of a state-of-the-art proportional directional valve in hydrostatic gear starting. It is shown that through a proper directional valve control algorithm one can shape the pressure during the starting of the gear, reducing in this way the accompanying system noise. The second part describes the digital fluid power (DFP) technology. Its main advantage is the system's simple design. One type of valve in 2 or 3 different sizes is sufficient to perform all the functions. In comparison with the often-used proportional valve, such a valve (called a binary valve), is characterized by simple circuitry, the absence of a position sensor, greater tolerance and lower sensitivity to impurities. An exemplary implementation of the DFP system as an additional drive in an off-road vehicle is presented. The third part describes the kinetostatic method for determining hydraulic system control parameters in order to increase the energy or environmental performance of the machine through the stabilization of the primary energy source and the selection of proper recuperation system components.

1. Introduction

Besides ensuring basic parameters, such as power output, driven element speed range, efficiency, etc., a drive system designer must also ensure proper dynamic properties specific to the machine being designed. Recently new-generation proportional directional control valves characterized by unprecedented properties have appeared on the market and are increasingly often used. The new directional valves achieve these properties owing to the novel way of driving the spool. In the patented design solutions the directional valve's spool interacts with a movable coil, instead of (as previously) with an armature. Thanks to this solution the inertia of the moving components can be reduced, whereby the cut-off frequency of the directional control valve can be increased up to the values so far characterizing only two-stage electrohydraulic amplifiers. This way of driving the directional control valve spool is called Voice Coil Drive (VCD) [1]. When designing drive systems, one must also consider solutions which will limit the emission of noise by the operating machines since hydrostatic drive system have, besides their commonly known advantages, a major drawback – they are a source of noise, sometimes exceeding the valid noise standards. The designer must also meet the additional requirements stemming from the specific application of the system, e.g. in special-purpose vehicles. The design of a special-purpose vehicle must overcome two main difficulties. One of them is the ability to negotiate nearly any terrain, but not at the expense of an overcomplicated vehicle structure. This entails the other requirement – the vehicle should be effective enough for its practical use to be profitable. For this reason it is necessary to employ an internal-combustion engine as the primary generator. Other solutions, e.g. electric motors, must be rejected since they necessitate the use of batteries with proper capacity, which are too heavy.

A hydrostatic transmission gear is one of the solutions very often used in agriculture and forestry. Its advantage is easy manoeuvrability making a continuous gear ratio within a certain range and gear change possible, which is useful in vehicles moving in steep terrain. Hence hydraulic systems based on the Digital Fluid Power (DFP) technology are increasingly often used in such applications.

In a DFP system the input and output parameters and the ones involved in the storage, transmission and processing of information can assume only specific values from a given range. There are two types of DFP systems: 1) based on the parallel connection and 2) based on the switching technology.

In each case, one should bear in mind the system operating parameters optimal for the adopted criterion. For this purpose one can use, e.g., the kinetostatic method of selecting operating parameters for a multisource drive system.

2. Use of proportional control technology for shaping hydrostatic gear starting

The throttling method and the volumetric method are the principle methods used to control the hydrostatic drive [2]. One of the reasons for the use of the proportional directional valve in the series throttling method to throttle the flow feeding the receiver is that thanks to this solution a change in the direction of the motion of the hydraulic receiver can be simultaneously effected. Figure 1 shows an exemplary schematic of a series throttling system for a hydrostatic gear with a proportionally controlled valve.

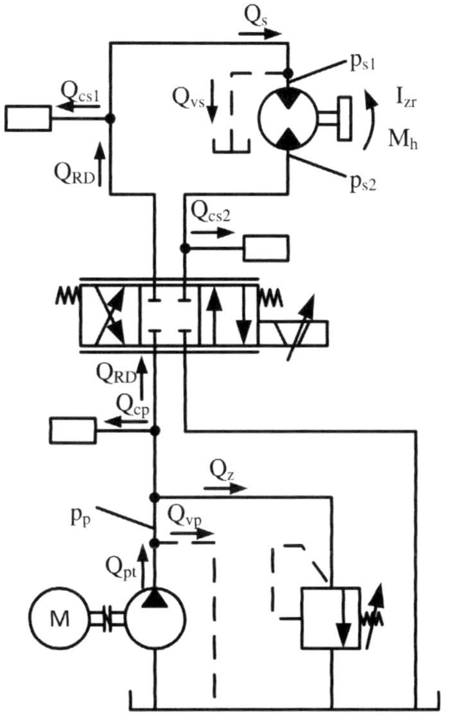

Figure 1. Schematic of series hydraulic system throttling hydrostatic gear: Q_{pt} – theoretical pump capacity 1, Q_{vp} – pump leakage losses, Q_z – flow through relief valve 3, Q_{RD}' – flow through proportional valve, Q_{Cp} – flow generated by compressibility within system volume between pump and valve controlling motion direction, Q_S – flow to hydraulic actuator, Q_{vs} – hydrostatic motor leakage losses, Q_{Cs1} – flow generated by compressibility within system volume between proportional directional valve and hydrostatic motor, Q_{Cs2} – flow generated by compressibility within system volume between hydrostatic motor and proportional directional valve.

IOP Conf. Series: Materials Science and Engineering **679** (2019) 012016 doi:10.1088/1757-899X/679/1/012016

A mathematical model of the series throttling control of a hydrostatic gear consists of a system of two equations: (1) for the continuity of the flow in the particular points of the hydraulic system and (2) for the equilibrium of hydrostatic motor shaft torques [3]. In order to solve the system of equations initial conditions must be defined [4]. Using the symbols from Figure 1 one can write the following equation of the continuity of the flow in the discharge flange of the pump:

$$Q_{pt} = Q_{vp} + Q_{cp} + Q_{RD} + Q_z \tag{1}$$

The balance of the rate of the flow through the throttling directional control valve has the form:

$$Q_{RD} = Q_s + Q_{sv} + Q_{cs1} \tag{2}$$

If the hydraulic losses in the conduits are omitted, the rate of the flow through the proportional directional control valve is described by the equation:

$$Q_{RD} = G_{RD}(t)\sqrt{p_p - p_{s1}} \tag{3}$$

In the considered system the dependence between the surface area through which the fluid flows in the proportional directional valve and the displacement of the spool is assumed to be linear. The dynamics of the proportional control valve in this system are described by the 1st order differential equation identical with the inertial term:

$$s(t)\cdot G_{RD\max} = G_{RD}(t) + T_{RD}\cdot\frac{dG_{RD}(t)}{dt} \tag{4}$$

where:
$G_{RD}(t)$ – the conductivity of the directional control valve, G_{RDmax} – the maximum conductivity of the directional control valve, T_{RD} – the time constant of the directional control valve, s(t) – a control parameter.

The equilibrium of the proportional valve pressure drops is assumed:

$$p_p - p_{s1} = p_{s2} \tag{5}$$

The flow towards the hydrostatic motor has the form:

$$Q_S = q_S \omega_S(t) \tag{6}$$

The flow induced by the compressibility of the working fluid and by the deformations of the components between the pump and the proportional control valve is written as:

$$Q_{cp} = c_p \cdot \frac{dp_p(t)}{dt} \tag{7}$$

where:
c_p – the capacitance of the fluid and of the conduits between the proportional control valve and the hydrostatic motor.

Between the proportional control valve and the motor the flow induced by the compressibility of the working fluid and by the deformations of the components has the form:

$$Q_{cs1} = c_{s1} \cdot \frac{dp_{s1}(t)}{dt} \tag{8}$$

where:
c_{s1} – the capacitance of the fluid and of the conduits between the proportional valve and the hydrostatic motor.

Between the motor and the proportional control valve the flow generated by the compressibility of the working fluid and by the deformations of the components is expressed as:

$$Q_{cs2} = c_{s2} \cdot \frac{dp_{s2}(t)}{dt} \tag{9}$$

where:
c_{s2} – the capacitance of the fluid and of the conduits between the motor and the proportional control valve.

The losses due to motor leakages are defined by the relation:

$$Q_{vs} = a_{vs} \cdot (p_{s1} - p_{s2}) \tag{10}$$

where:
a_{vs} – a motor leakage coefficient, p_{s1}, p_{s2} – the pressure in the flanges of the hydrostatic motor.

The flow through the relief valve [4, 5], amounts to:

$$Q_z(t) + T_z \cdot \frac{dQ_z(t)}{dt} = h_z \cdot \left[p_p(t) - p_0 \right] \text{ for } p_p(t) > p_0 \text{ and } Q_z(t) = 0 \text{ for } p_p(t) \leq p_o \tag{11}$$

where:
T_Z – the time constant of the relief valve, h_z – a valve amplification factor, po – the relief valve opening pressure.

The hydrostatic motor shaft torques equilibrium condition is expressed by the relation:

$$q_s \cdot (p_{s1} - p_{s2}) = M_h + f \cdot \omega_s(t) + I_{zr} \cdot \frac{d\omega_s(t)}{dt} \tag{12}$$

where:
p_{s1} – the hydrostatic motor suction flange pressure, p_{s2} – the hydrostatic motor discharge flange pressure, q_s – the (geometric) specific absorbing capacity of the motor, f – the (experimentally determined) viscous friction coefficient [5], M_h – the motor shaft braking torque, ω_s – the angular speed of the hydrostatic motor, I_{zr} – the reduced to the motor shaft solid moment of inertia of the rotating masses.

The initial conditions are:

$$p_p(0) = p_0; \; p_s(0) = 0 \tag{13}$$

The boundary condition is defined as follows:

$$\text{when } q_s(p_{s_1} - p_{s2}) \leq M_h, \text{ then } \omega_s = 0 \text{ and } (d\omega_s)/(dt) = 0 \tag{14}$$

After the mathematical model equations are parametrized and the initial conditions assumed, the model can be numerically solved and the time history diagrams of the pressure in the motor suction flange can be graphically presented (Figure 2).

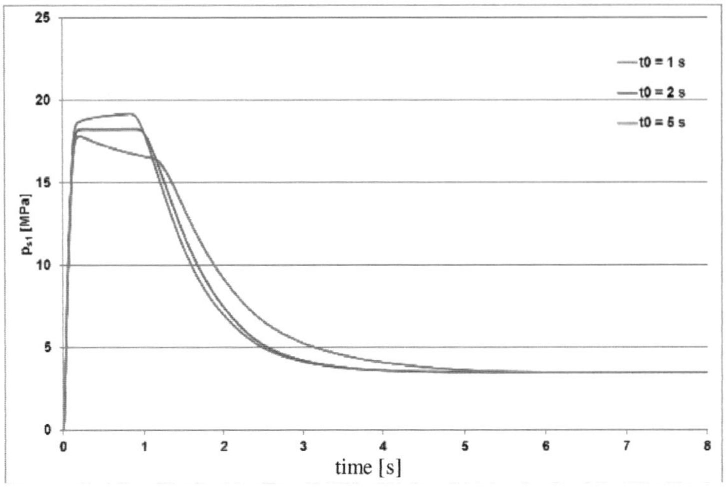

Figure 2. Pressure p_{s1} in motor suction flange of hydrostatic gear with proportional directional control valve incorporated in series versus time, assuming directional control valve resetting time as parameter.

A change in the control signal rise time causes a change in the maximum pressure value in the suction flange of the motor. The graphically presented solutions are for different values of control signal rise time t_0: 1 s, 2 s, 5 s. The change in the control signal fed to the proportional directional control valve's coils over time ($s(t)$) is shown in Figure 3 and described by the relation:

$$s(t) = \left(\frac{i_0}{i_{\max}} + \frac{1 - \frac{i_0}{i_{\max}}}{t_0} \cdot t \right) \quad (15)$$

where:
i_0 – the minimum value of the control current, i_{max} – the maximum value of the control current, t_0 – the control signal rise time.

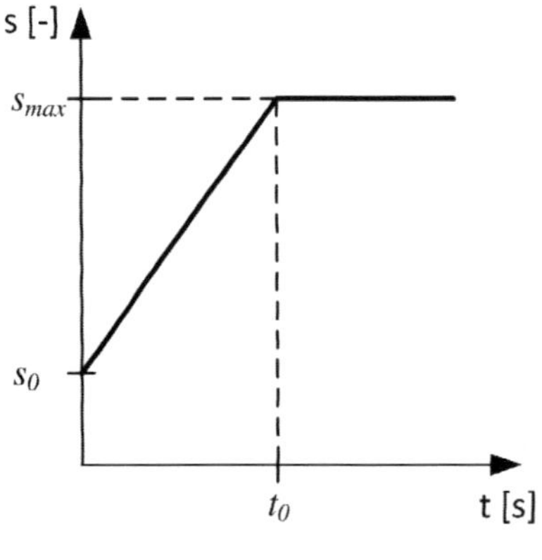

Figure 3. Graph of signal controlling proportional directional control valve, to – control signal rise time during series throttling control.

During the starting of the gear by means of the series throttling control method the positive displacement pump is loaded with the relief valve set point regardless of the rise time of the signal controlling the proportional control valve. But one can have an influence on the duration of this load – the faster is the valve reset, the shorter the duration of the pump load resulting from the operation of

IOP Conf. Series: Materials Science and Engineering **679** (2019) 012016 doi:10.1088/1757-899X/679/1/012016

the overflow valve. The proportional control valve resetting time affects the hydraulic motor feed pressure – mainly the maximum value of pressure p_{s1} during starting. By reducing the maximum pressure values during the starting of the gear one can reduce the latter's operational noisiness in this phase of motion [5].

3. Special-purpose vehicle design incorporating hydraulic components controlling system output power

A parallelly connected DFP system includes many connections and the output effect is controlled through a change in the states of the combinations of the components. Such a system has a specific number of discrete output values and there is no need to employ switches. Figure 4a shows an assembly of valves connected in parallel, performing the function of a two-way valve. Such a structure is called a Digital Flow Control Unit (DFCU).

Since the valves in a DCFU can be only completely opened or completely closed, the rate of flow is adjusted in discrete steps, as opposed to the proportional valve in which proportioning is continuously adjusted.

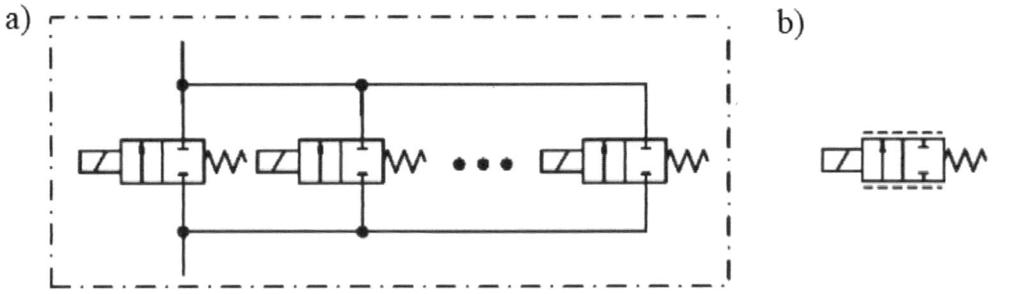

Figure 4. Digital flow control unit: a) schematic, b) symbol [6].

Whereas the switching technology uses the fast and continuous switching of one or several elements and the output is adjusted through, e.g., pulse width. The symbol representing such a valve is shown in Figure 5b. This is one of the commonest ways of controlling a medium flow area. Theoretically the latter can assume any value, but the valve's dynamics limit boundary work cycle values. The controllability also depends on the switching frequency. A low switching frequency improves the control of a medium flow area, but it contributes to pressure fluctuations. Usually it is necessary to use damping elements in order to reduce noise.

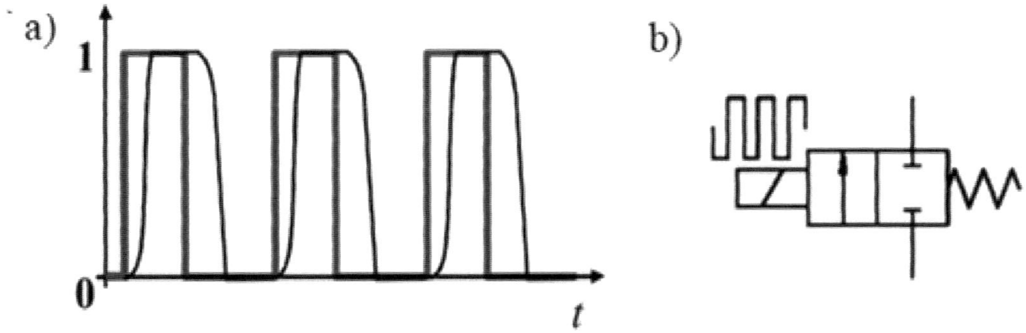

Figure 5. On/off valve controlled by pulse width: a) signal graph (blue curve) and valve loading graph (black curve), b) valve symbol (based on [6, 7]).

The DFP technology also involves the use of digital pumps (Figure 6). In a conventional pump the transport of a liquid is effected through cam disk deflection, which is set for the pump delivery best for the particular operating conditions. In the switching technology, the flow from a constant capacity pump is controlled by means to the on/off valve mentioned above. In the case of a parallel connection, the pump actually consists of several constant capacity pumps mounted on a single shaft.

IOP Conf. Series: Materials Science and Engineering **679** (2019) 012016 doi:10.1088/1757-899X/679/1/012016

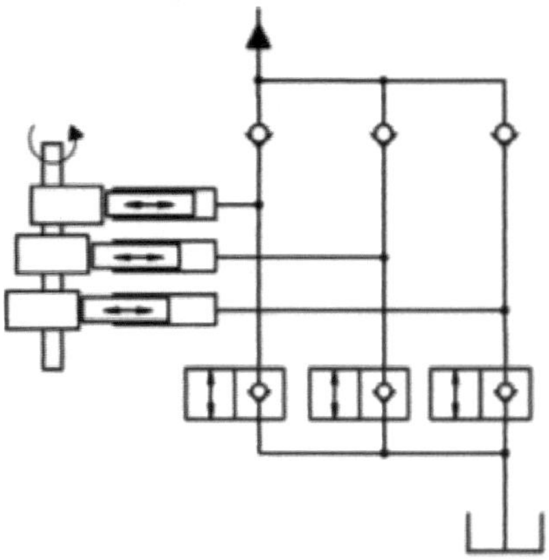

Figure 6. Principle of digital pump operation with three pistons [8].

Figure 7 shows a design of a hydraulic drive incorporating digital technology elements and Figure 8 - schematic of drive using powersplit technology. The engagement and disengagement of the drives (a conventional drive and a hydrostatic drive) in a special-purpose vehicle will be effected by the powersplit gear used in electrical or hydraulic hybrids.

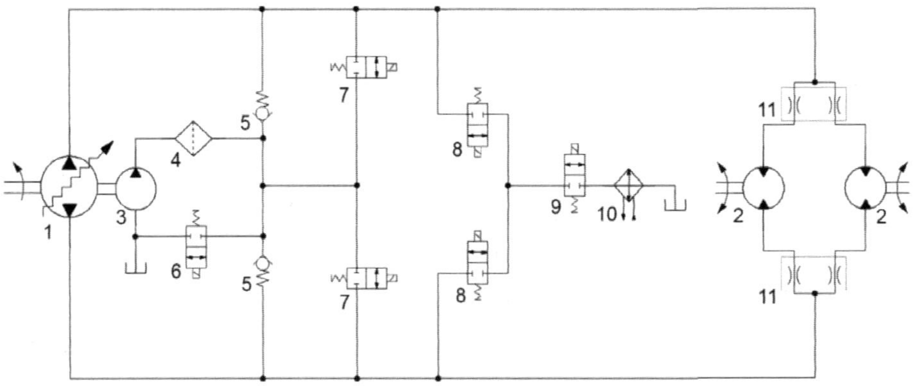

Figure 7. Hydraulic diagram of selected solution.

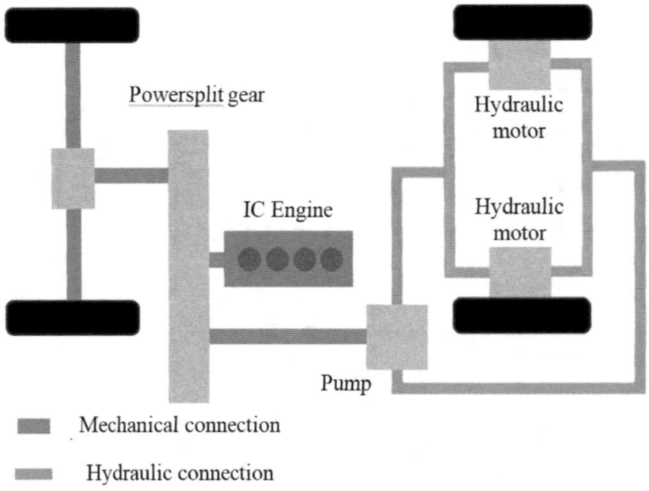

Figure 8. Schematic of drive using powersplit technology.

The solutions described above encounter difficulties in practice, mainly due to the small selection of digital hydraulics components on the Polish market. Since only one digital pump was commercially available the original design had to be somewhat modified, but the modification did not infringe the project's guidelines. The next step is to develop a control program, which when properly designed will increase the operating efficiency of the vehicle.

4. Kinetostatic method of selecting operating parameters for multisource hydrostatic drive system

This method is used to determine – for a multisource drive system built of components with known characteristics, a load in the form of a machine duty cycle and an adopted criterion – the following parameters:
- the optimal primary energy source working point,
- the initial parameters of the secondary energy source,
- the modes of controlling the other drive system components.

This is done under the following assumptions:
- the primary energy source is to work in one point throughout the duty cycle of the device,
- the secondary energy source charge level is to be the same the beginning and end of the duty cycle.

The method yields a set of parameters for controlling the multisource drive system for the adopted criterion.

The method of selecting the operating parameters for a multisource drive system is based on the refined kinetostatic method [9, 10, 11, 12], shown in schematic way on Figure 9.

Step 1. On the motor performance characteristic define the possible allowable operation area. On this basis determine the considered set of engine operation points M_{sp}, ω_{sp}, G_{cj}, CO, NO_X and exhaust gas opacity.

Step 2. Present, in the form of a table or in an analytical form, the machine load characteristic in the duty cycle of duration T. The characteristic comprises dependences between load (force) torque $M_o(t_j)$ and (linear) angular velocity $\omega_o(t_j)$. Assume time increment Δt values. Select energy accumulator parameters: capacity and initial charge level $E_a(t_i=0)$.

Step 3. Compare the values of internal-combustion engine power $N_{spj} = f(M_{spj}, \omega_{spj})$ for the adopted j-th allowable area point with power demand $N_{oi}(M_{oi}, \omega_{oi})$ over time t_i, consistently with the load cycle.

Since:

- the primary source power and the load power are expressed by the respective values of their stress (torque) variables M_{spj} and M_{oi} and intensity variables ω_{spj}, ω_{oi} and
- in the general case, the values of engine power N_{spj} and load N_{oi} are different,

the task of the hydrostatic power transmission system is to:
- appropriately transform the internal-combustion engine's stress variable M_{sp} and intensity variable ω_{sp} to the form required by the load's stress variable M_o and intensity variable ω_o,
- balance the power supplied by the primary energy source and the load power demand by carrying off the excess energy to the accumulator or making up for an energy shortage by drawing energy from the accumulator.

The primary source power in the form of signals M_{sp} and ω_{sp} is transmitted to the shaft of a displacement pump with a known capacity. For given pressure p input quantities M_p and ω_p are transformed, by solving the equations describing actual pump energy characteristics M_p and Q_p, into discharge flow rate controlling element tilt (ε_p) signals.

The instantaneous value of the load power in the form of signals M_o and ω_o is transmitted to the shaft of the hydrostatic motor with a variable absorbing capacity. For given pressure p quantities M_o and ω_o are transformed, by solving the equations describing the hydrostatic motor's actual energy characteristics M_s and Q_{sh}, into motor controlling signal ε_{sh}.

IOP Conf. Series: Materials Science and Engineering **679** (2019) 012016 doi:10.1088/1757-899X/679/1/012016

The energy losses connected with flows Q_p and Q_{sh} in the hydraulic system, relative to the summing node in which pressure p prevails, are mapped in the next blocks. At the given value of pressure p power balancing takes place in the summing node through the satisfaction of the flow continuity equation: $Q_p + Q_{sh} + Q_a = 0$. In the case of primary source excess power relative to the load, the accumulator draws a liquid stream with power $p \cdot (Q_p - Q_{sh})$, while in the case a shortage, it gives up power $p \cdot (Q_{sh} - Q_p)$.

Pressure p in the summing node is closely connected with the energy state of the accumulator. However, in the general case, this is not a simple and unequivocal relationship. It depends on the accumulator used and the way of controlling it. In the considered case, a hydraulic accumulator, incorporated directly into the hydraulic system, was used. The algorithm for this solution is shown in Figure 10.

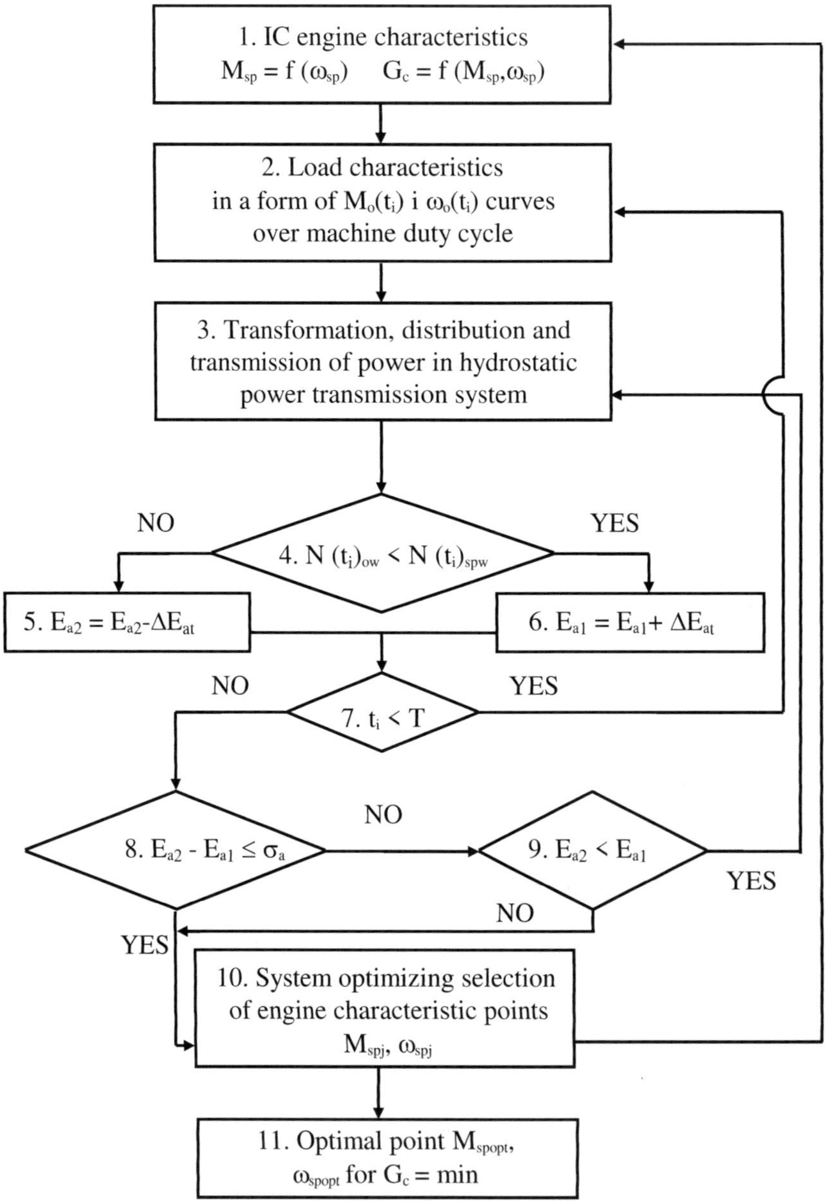

Figure 9. Procedure for determining control parameters for prescribed hybrid system load – refined kinetostatic method [11, 12].

IOP Conf. Series: Materials Science and Engineering **679** (2019) 012016 doi:10.1088/1757-899X/679/1/012016

In **block A** pressure p in the summing node is calculated for the given iteration (*k-th*). The pressure value depends on the accumulator charge status. Then the values of hydrostatic unit control parameters ε_p, ε_{sh} are calculated for the given load M_o and internal-combustion engine torque M_{spj} (**block B**). The values, relative to the respective previous values, over time t_i -Δt must satisfy the limit resetting speed condition.

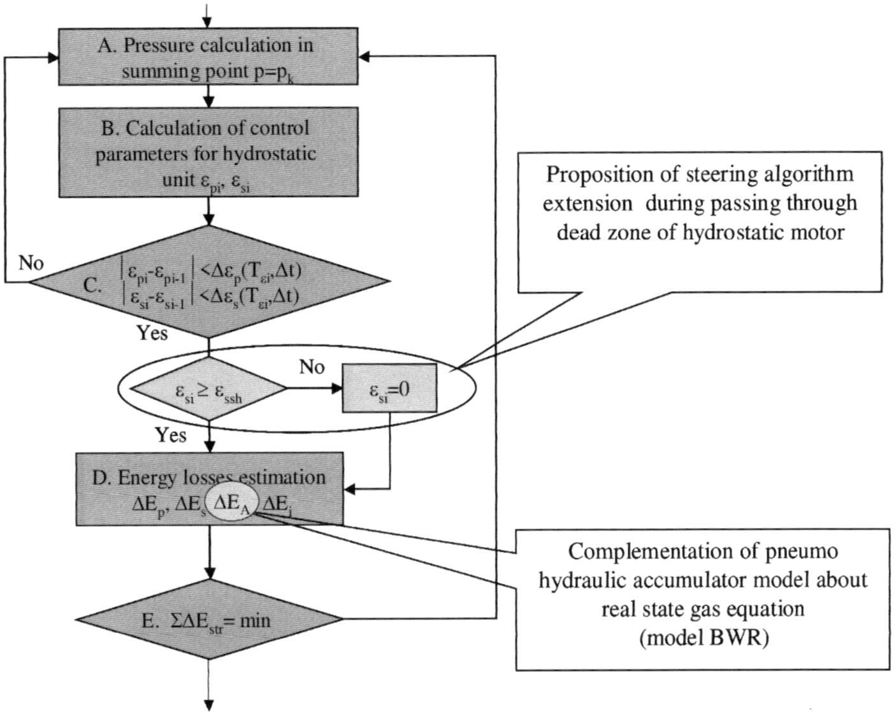

Figure 10. Development and refinement of algorithm in block 3 dealing with power transformation and transmission in hydrostatic power transmission system [11, 12].

If the conditions in **block C** are not satisfied, go back to block A. If the maximum pump and hydrostatic motor resetting speeds are not viable, check whether the tilt of the hydrostatic motor disk is larger than the so-called hydrostatic motor self-braking interval. If this condition is satisfied, go to **block D**. Since the motor will find itself in the zone mentioned above, reduce disk tilt to 0 in the shortest possible time to minimize volumetric losses. In **block D** calculate the energy losses. First determine the real pump capacities Q_p and motor absorbing capacities Q_{sh}.

Then determine the energy losses over time t_i - t_{i-1} = Δt in:

- the displacement pump – $\Delta E_p(\omega_p, p_p, \Delta_p)$,
- the hydrostatic motor – $\Delta E_{sh}(\omega_{sh}, p_{sh}, \Delta_{sh})$,
- the hydraulic system –$\Delta E_t(Q_{sh}, Q_p)$,
- in the energy storage system – $\Delta E_a(E_a, N_a)$ (using the Benedict Rubin Weber (BWR) model).

The power losses in the energy storage system depend on the accumulator charge status E_a, charging and discharging rates $\frac{dE_a}{dt}$, i.e. the supplied or drawn power, the duration of energy storage in the accumulator and the latter's structural parameters: the wall thickness, the thermal conductivity coefficient of the walls, the ambient temperature, etc.

In **block E** decisions about going to the next (4th) step of the algorithm or the next ($k+1$) iteration and a search for the next converter control are made. Thus the main problem consists in selecting such energy converter controlling signal E_a that for:

- internal-combustion engine working point M_{spj}, ω_{spj} and
- load characteristic point $M_o(t_i)$, $O_o(t_i)$ for instant t_i

energy loss sum $\Sigma\Delta E_{str}$ is minimal.

__Step 4.__ Compare the values of the instantaneous capacities reduced to the summing node, i.e. with energy transformation and transfer losses taken into account.

__Step 5.__ Calculate values E_{a2} of the energy abstracted from the accumulator.

__Step 6.__ Calculate the energy supplied to the accumulator over time $\Delta t = t_i - t_{i-1}$, and reduced to the summing node. The operation of step 5 or 6 depends on the sign of inequality $N(ti)_{ow} < N(t_i)_{spw}$.

__Step 7.__ Stop the calculations for given cycle instant t_i; when $t_i < T$, repeat the algorithm loop for $t_i = t_i + \Delta t$ until time $t_i = T$. When duty cycle $t_i = T$ ends, go to the next step.

__Step 8.__ Test condition $\left|E_{a2} - E_{a1}\right| < \sigma_a$. Depending on the result obtained for the given load cycle, do the following:

- if the condition specified in **step 8** is satisfied, select new internal-combustion engine working point M_{spj}, ω_{spj},
- if the condition is not satisfied, consider the following two options:

__Step 9.__

1. When for the given cycle the energy supplied to the accumulator is lower than the energy drawn from the latter, the energy balance of this cycle is negative, with the shortage exceeding σ_a. This option is unacceptable and so go to step 10 and select a new internal-combustion engine working point M_{spj}, ω_{spj}.
2. When for the given cycle the energy supplied to the accumulator is higher and the excess energy exceeds σ_a, go back to block 3 (Figure 10) concerning the energy converter control signal and repeat the load cycle calculations for the given internal-combustion working point M_{spj}, ω_{spj}.

__Step 10.__ The energy analysis is performed by the optimizing system until point M_{spopt}, ω_{spopt}, in which the quality criterion (fuel consumption G_c reaches the minimum) for the given cycle is satisfied, is found.

In the above method of selecting the parameters of a multisource hydrostatic drive system [9] one can adopt any control criterion, e.g. minimal fuel consumption G_e or minimal atmospheric harmful emission (CO, NO_x, smoke pollution). Consequently, one can select a criterion proper for the place of machine operation, e.g. on throughways and in non-residential area one can adopt the minimal cost (min. G_e) criterion, increasing thereby exhaust gas emission, whereas in mines and built-up areas the fuel cost is of secondary importance.

5. Conclusion

The exemplary state-of-the-art hydraulic drive solutions presented in this paper indicate that the main advances are made in the areas of: the hydraulic drive and the control of hydraulic components, the continuous (proportional) control technology and the digital fluid power (DFP) technology. As a result, hydraulic systems can achieve unprecedented performance and their efficiency can be increased through the use of an appropriate control algorithm. The costs and availability of hydraulic components put some limitations on the use of the new technologies of controlling them. Moreover, the working fluid cleanness requirements for proportional valves are higher.

The kinetostatic method described in this paper can be used not only to determine the drive system control optimal for the energy or environmental criteria adopted for a given machine duty cycle, but also to estimate the potential benefits from the use of multisource systems instead of the conventional ones. Thus, this method is a tool which can aid engineers in upgrading conventional drive systems and estimating the benefits gained by replacing a conventional drive system with a multisource one.

References

[1] *Katalog Parker Hannifin HY 14-2550/US*

[2] Stryczek S *Napęd hydrostatyczny*, Warszawa: WNT, 2003

[3] Palczak E *Dynamika elementów i układów hydraulicznych*, Wrocław: Ossolineum, 1999

[4] Tomczyk J *Modele dynamiczne elementów i układów napędowych hydrostatycznych*, Warszawa: WNT, 1999

[5] Kudźma Z *Tłumienie pulsacji ciśnienia i hałasu w układach hydraulicznych w stanach przejściowych i ustalonych*, Wrocław: Oficyna Wydawnicza Politechniki Wrocławskiej, 2012

[6] Linjama M "Digital Fluid Power - State of the art," in *The Twelfth Scandinavian International Conference on Fluid Power*, Tampere, 2011

[7] Linjama M "Digital Hydraulics" [Online]. Available: https://mycourses.aalto.fi/pluginfile.php/870043/mod_resource/content/1/Digital%20Hydraulics%20lecture%20slides.pdf.

[8] Rampen W and Salter H "The Digital Displacement Hydraulic Piston Pump" in *Proceedings of the 9th International Symposium on Fluid Power*, Cambridge

[9] Kędzia K "An algorithm for the determination of the control parameters of a multisource drive system" *Technical Transactions,* pp. 173-182, vol. 3 2017

[10] Kędzia K "Development of kinetostatic method algorithm of control system of multisources driving system" *Hidraulica,* pp. 39-45, 3/4 2009

[11] Kędzia K *Metoda optymalizacji energetycznej i ekologicznej hydrostatycznego wieloźródłowego układu napędowego,* Wrocław: Praca doktorska, 2004

[12] Kędzia K *Wyznacznik zmienności cyklu obciążenia wieloźródłowego hydrostatycznego układu napędowego*, Wrocław: Oficyna Wydawnicza Politechniki Wrocławskiej, 2014

IOP Conf. Series: Materials Science and Engineering **679** (2019) 012017 doi:10.1088/1757-899X/679/1/012017

The simulation of fluid flow in safety elements of longwall shield support hydraulic legs

S Rajwa[1]*, T Janoszek[1] and J Świątek[1]
[1] Central Mining Institute, Plac Gwarków 1, Katowice, Poland

*E-mail: jswiatek@gig.eu

Abstract. The paper presents the research on pressure variation in elements securing the shield support against overloading. The chosen constructions of connecting pipe were analysed and its flow characteristic were determined. The operational parameters of different connections were also considered in light of their interaction with whole hydraulic control system. The results allowed to determine of the optimal connecting pipe curve with regard to the pressure in whole pressure safety system.

1. Introduction

Longwall powered support intended for operation under the conditions of rock mass tremor hazards should be yielding, i.e. it should be able to take over dynamic loads [6, 8].

This characteristic is required according to the regulations included in the Ordinance of the Minister of Energy of 23 November 2016, Dz. U. No 2017, item 1118, para. 523, sec. 1, pt. 1. The evaluation of yielding supports as well as the guidelines for their construction are prepared at the Central Mining Institute according to own methodology. It is currently the only method of support yield evaluation that is applied broadly in practice. It is an analytical method utilising many years of operational experience and the technical parameters of the support units. This approach also constitutes a part of the support selection method for geological and mining conditions [5, 7, 12].

Its use is based on determining a safe operational range of a support unit for a given situation. Result verification of the application of the method in question is systematically conducted by GIG employees, including the use of increasingly more broadly utilised support monitoring systems [3, 4].

One of the key support unit safety elements defined in the GIG method is the fluid flow in the system securing the legs from overloading. Such a system is comprised of a pressure relief valve and a pipe supplying it with fluid from the leg working space. The pipe flow characteristics, understood as the relation between the pressure and the fluid stream volumetric flow, and the valve parameters determine the efficiency of securing the leg from overloading. Deteriorating mining conditions (increasing depths of the extracted deposits, rock mass tremor hazards) result in the necessity to increase the support unit load capacity by utilising legs with diameters $\varnothing \geq 0.32$ m. Additionally, the current trend of placing the leg control systems as high as possible above the floor considerably extends the distance needed for the fluid stream to flow from the leg working space to the pressure

IOP Conf. Series: Materials Science and Engineering **679** (2019) 012017 doi:10.1088/1757-899X/679/1/012017

relief valve. The flows required to secure the high-diameter legs necessitate the optimisation of the fluid stream, directed at decreasing the losses in continuous and local flows.

The subject of this publication is an example use of numerical analysis using the SolidWorks Flow Simulation software to optimise the system securing the hydraulic leg from overloading, and particularly to optimise the shape of the fluid stream supplied by a connecting pipe from the leg working space to the pressure relief valve.

2. Leg securing system flow characteristics

The leg securing system is characterised by parameters that express continuous flow as a set of relationships with pressure at the system entry. These are non-linear relationships dependent on multiple factors, such as: fluid type (mono- or diphase), local and continuous losses (shape and geometry of the fluid stream, roughness, Reynolds number), characteristics of the fluid flow through the valve. Fluid leaks can be continuous or undergo periodic variations as a result of the two-state valve operation, i.e. open/closed. This publication presents issues related only to the steady continuous flow. The hydraulic system securing the leg from overloading is presented as a diagram in Figure 1a. The fluid stream is drained from the secured leg space via a port, usually a steel pipe with an outlet. The flow of the entire system can be described as a sum of the characteristics of the connector and the valve, presented as an exponential polynomial. This form makes it possible to sum the relationships for a series connection of system elements where the flow for individual elements is identical – Figure 1b [10].

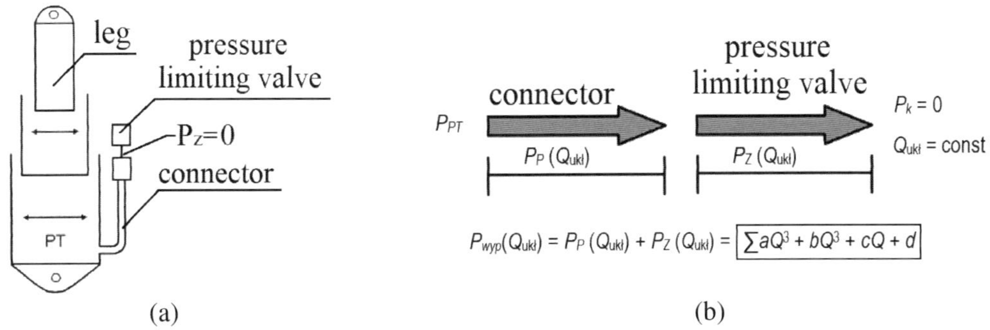

Figure 1. Flow characteristics of the system securing the leg from overloading and its determination method [10]; a – general diagram, b – flow computational relationships. Where: P_{PT} – pressure under the piston, P_P – pressure drop in the connector for a specified mass flow, Q_{ukl} – system volumetric flow, P_Z – pressure drop in the valve for a specified mass flow, P_K – pressure at the valve exit, Q – volumetric flow.

3. Scope and description of the conducted leg port analyses

The first stage of the numerical modelling with the use of CFD (Computational Fluid Dynamics) methods is to define its purpose, i.e. to consider the physical phenomena, the conditions of the uniqueness of the numerical model solution, and the scale of the computational area. The second stage is to develop the geometry of the structure that will constitute a reflection of the actual subject of study. Once the geometric model is built, the next stage is discretisation, i.e. generating a computational grid that will constitute the area of the numerical solution of the analysed problem. The final stage of the modelling process is to define the correct settings for the analysed problem (the solver) and to perform numerical calculations with systematic monitoring of the convergences of the obtained numerical solution.

The analysed powered support unit leg port is presented in Figure 2. It is comprised of a specifically shaped pipe and a special tip for connecting the relief pressure valve. Both the elements are made from low-alloyed constructional steel, which is characterised by high strength and ductility, good treatment and fine-grained alloy structure.

IOP Conf. Series: Materials Science and Engineering **679** (2019) 012017 doi:10.1088/1757-899X/679/1/012017

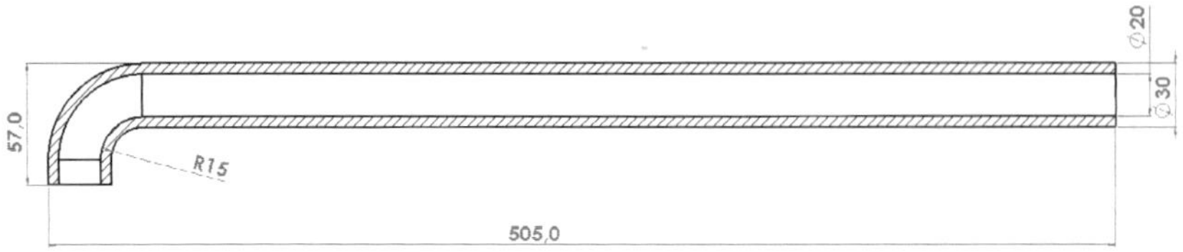

(a)

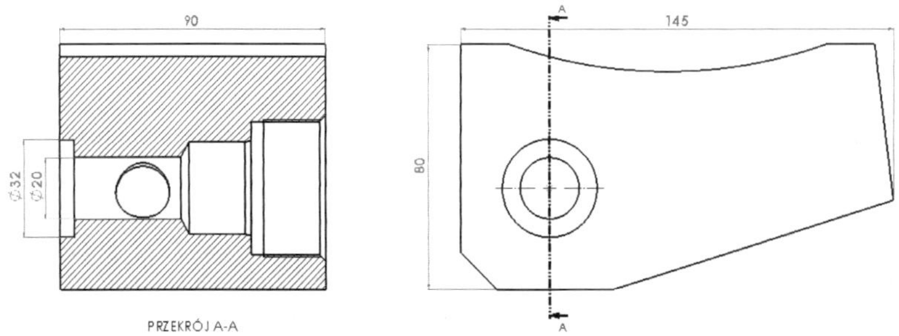

(b)

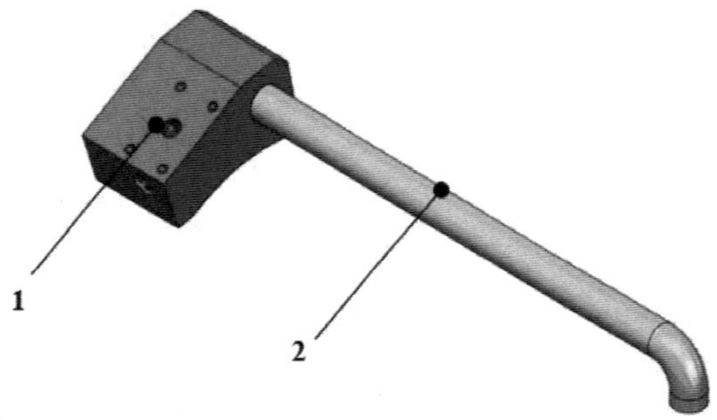

(c)

Figure 2. Leg securing system. a - connector cross section, b - limiting pressure valve block, c - analysed longwall powered support leg securing system assembly model, where: 1 – limiting securing system block, 2 – connector.

The main goal of the analysis was the volumetric fluid flow through the analysed connector as a function of pressure. The following calculation variants were adopted:

influence of the connector geometry on the pressure characteristics – variant I,

influence of the connector cross-section on the pressure characteristics – variant II,

influence of the air content in the fluid on the pressure characteristics – variant III.

A discretisation area was generated based on the developed connector assembly model in SolidWorks software, in the form of a numerical grid composed of 430535 fluid cells, where 87651 fluid cells contacting solids, which represents geometry occupied by the fluid – Fgure 3 (volume of fluid is 0.000198 m^3).

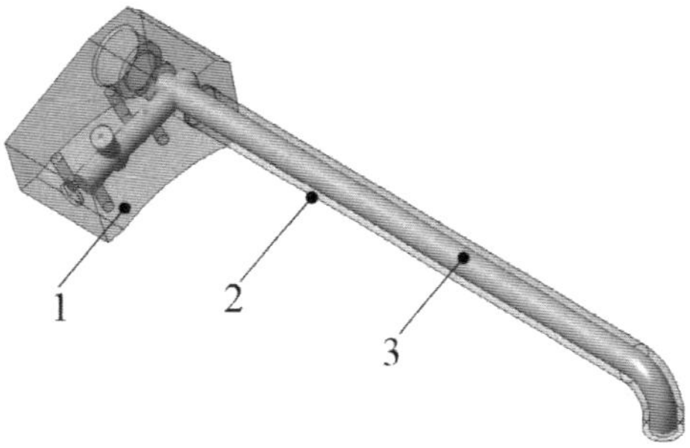

Figure 3. Leg securing system: 1 – pressure limiting valve block, 2 – connector, 3 –fluid volume.

The following boundary conditions were included for the purposes of the model:
1. For variants I and II, the calculations involved fluid in the form of pure water with the following parameters:
density ρ = 999.76 [kg·m^{-3}],
dynamic viscosity μ = 0.0017912 [Pa·s],
specific heat c_p = 4219.9 [J·kg^{-1}·K^{-1}],
volumetric flow variation range V = 0÷2000 dm^3·min^{-1},
thermal conductivity coefficient λ = 0.56104 [W·m^{-1}·K^{-1}].
2. For variant III, the calculations involved fluid in the form of a mixture of water and air, with the following air parameters:
density ρ = 1.2 [kg·m^{-3}],
dynamic viscosity μ = 6.04e^{-6} [Pa·s],
specific heat c_p = 1029.1 [J·kg^{-1}·K^{-1}],
volumetric flow variation range V = 0÷2000 dm^3·min^{-1},
thermal conductivity coefficient λ = 0.00779 [W·m^{-1}·K^{-1}].

The results of the conducted modelling are presented on charts as a relationship of P=f(Q), and additionally in the form of an exponential polynomial for a selected case.

4. Numerical analysis results

The analyses of individual leg securing system variants were conducted using the *SolidWorks 2018* software with the *Flow Simulation* tool. *SolidWorks* is a program for computer-aided design, used for constructing 2D and 3D models. This simple very efficient computer program allows designers to create very detailed elements and assemblies at the design stage as well as working drawings. *SolidWorks* is an engineering tool that can support multiple stages of the development of a given product. It makes it possible to inspect a design directly before production by means of a broad range of available tools, from computational fluid dynamics (CFD) to the analysis of static and dynamic strength of a studied prototype construction, based on the finite element method (FEM) [2, 9].

The purpose of the fluid flow process simulation supported by CFD methods in the space that constitutes the volume occupied by the analysed fluid is to obtain a solution to a set of differential equations which interpret the law of conservation of mass and momentum of the moving fluid (Navier-Stokes equation). These fundamental equations, which express the behaviour of a fluid flow through a given geometry in *SolidWorks Flow Simulation*, are relationships defined in the following forms [2, 9]:
- mass conservation equation:

$$\frac{\partial p}{\partial t} + \nabla(p v) = 0 \tag{1}$$

- Navier-Stokes equation:

$$\rho \frac{\partial v}{\partial t} = -\nabla p + \rho g + \mu \nabla^2 v \tag{2}$$

where:

ρ – fluid density [kg·m^{-3}],
υ – fluid velocity [m·s^{-1}],
p – fluid pressure [Pa],
μ – fluid dynamic viscosity [Pa·s].

The influence of disturbances in the fluid transfer process within the given geometry were interpreted with a ***k-ε*** turbulence model. Solving this model comes down to determining the value of turbulent viscosity $\boldsymbol{\mu_t}$ using turbulence kinetic energy $\boldsymbol{k}$ and dissipation rate ε related to the energy dissipation resulting from the occurrence of internal resistances to motion of a fluid flow through a channel. The turbulent viscosity $\boldsymbol{\mu_t}$ model of the fluid is expressed using an equation defined in the following form in SolidWorks Flow Simulation:

$$\mu_t = \rho C_\mu \frac{k^2}{\varepsilon} \tag{3}$$

Fluid transport equations for turbulence kinetic energy k and dissipation ε are expressed in SolidWorks Flow Simulation as relationships in the following forms [2, 9]:

— for turbulence kinetic energy:

$$\frac{\partial \rho k}{\partial t} + \frac{\partial \rho k v_i}{\partial x_i} = \frac{\partial}{\partial x_i}\left(\left(\mu + \frac{\mu_i}{\sigma_k}\right)\frac{\partial k}{\partial x_i}\right) + \tau_{ij}^R \frac{\partial v_i}{\partial x_j} - \rho\varepsilon + \mu_t P_B \tag{4}$$

— for dissipation energy:

$$\frac{\partial \rho k}{\partial t} + \frac{\partial \rho \varepsilon v_i}{\partial x_i} = \frac{\partial}{\partial x_i}\left(\left(\mu + \frac{\mu_i}{\sigma_k}\right)\frac{\partial \varepsilon}{\partial x_i}\right) + C_{\varepsilon 1}\frac{\varepsilon}{k}\left(f_1 \tau_{ij}^R \frac{\partial v_i}{\partial x_j} + C_B \mu_t P_B\right) - f_2 C_{\varepsilon_2}\frac{\rho \varepsilon^2}{k} \tag{5}$$

where:
$C_{\varepsilon 1}$ – empirical constant, $C_{\varepsilon 1} = 1.44$,
$C_{\varepsilon 2}$ – empirical constant, $C_{\varepsilon 2} = 1.92$,
C_μ – empirical constant, $C_\mu = 0.09$,
k – velocity fluctuation (turbulence) kinetic energy [m^2·s^{-2}],
P – local vorticity fluctuation production,
ε – turbulence kinetic energy dissipation rate [m^2·s^{-3}],
μt – turbulent viscosity [Pa·s],
σk – turbulent Prandtl number $\sigma_k = 1.0$,
$\sigma \varepsilon$ – turbulent Prandtl number $\sigma_\varepsilon = 1.3$,
ρ – fluid density [kg·m^{-3}].

4.1. Stream geometry influence

In order to demonstrate the influence of the variations in fluid geometry on the pressure distribution along the analysed leg securing system (Figure 2), numerical simulations were performed for the model, where the bend radius of connector was modified from the value of R=15 mm (Figure 2a) to a value of R=40 mm (Figure 4a). Figure 4 presents the basic dimensions of the modified connector (Figure 4a) and the view of the spatial model of the leg securing system in the assembly (Figure 4b).

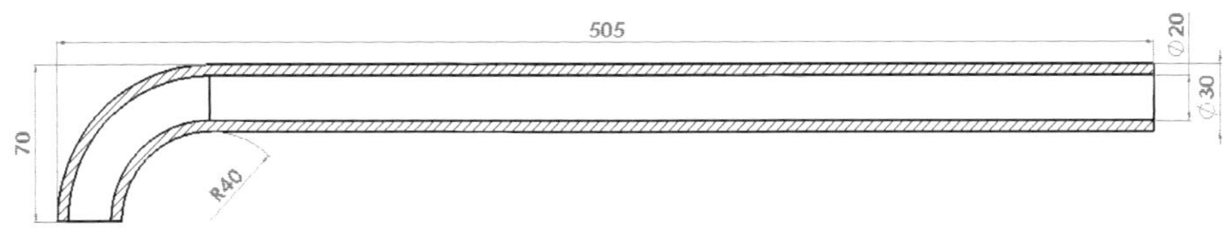

(a)

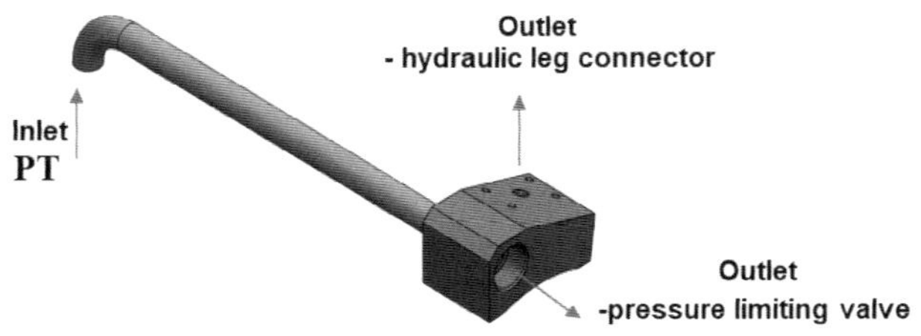

(b)

Figure 4. View of the modified powered support leg securing system: a – connector cross section, b – spatial view.

The numerical calculations were conducted within the scope of water volumentric flow variations at a range of 0 to 2000 $dm^3 \cdot min^{-1}$, simulating the fluid supply to the modified (Figure 4) and unmodified (Figure 2) leg securing system, as depicted in Figure 4. Total pressure variation measurements were performed at the outlet of the pressure limiting valve.
The obtained results of the conducted numerical simulations are presented in Figure 5 in the form of a P=f(Q) chart, where P is the total pressure distribution variation along the powered support leg securing system.

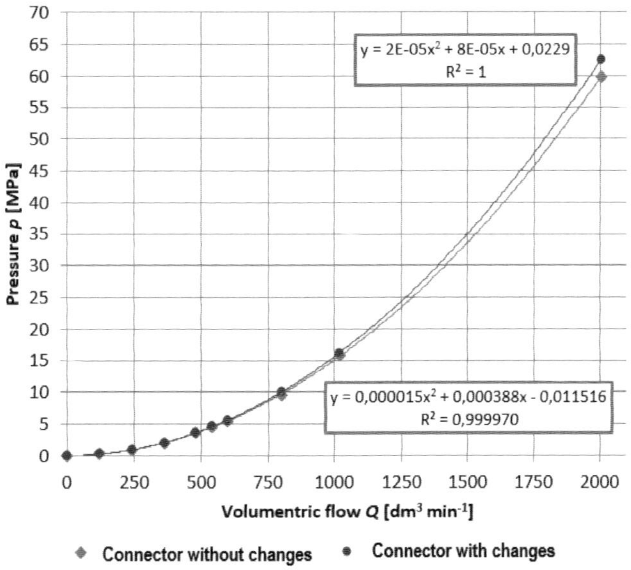

Figure 5. Characteristics of the total pressure variation at the powered support unit leg securing system outlet depending on the fluid volumetric flow.

IOP Conf. Series: Materials Science and Engineering **679** (2019) 012017 doi:10.1088/1757-899X/679/1/012017

Additionally, maps of the total pressure variation distribution along the studied powered support unit leg port geometries for a fluid volumetric flow of 1000 $dm^3 \cdot min^{-1}$ are presented in Figure 6 as well as for a fluid volumetric flow of 2000 $dm^3 \cdot min^{-1}$ in Figure 7.

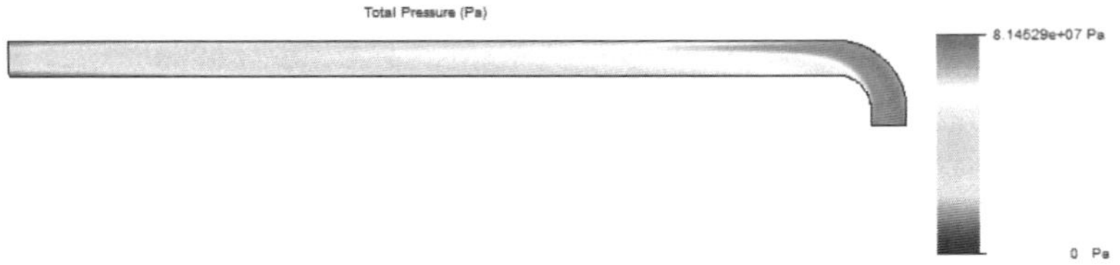

(a)

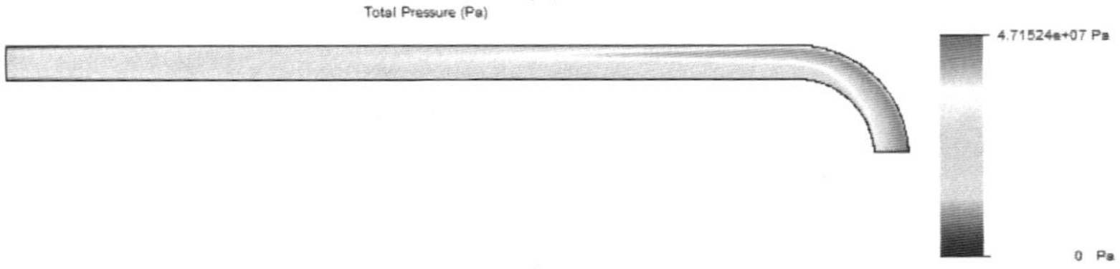

(b)

Figure 6. Maps of the total pressure variation distribution along the studied powered support leg securing system for a fluid volumetric flow of 1000 $dm^3 \cdot min^{-1}$: a – pressure variation distribution for a powered support leg securing system geometry from Figure 2, b – pressure variation distribution for a powered support leg securing system geometry from Figure 4.

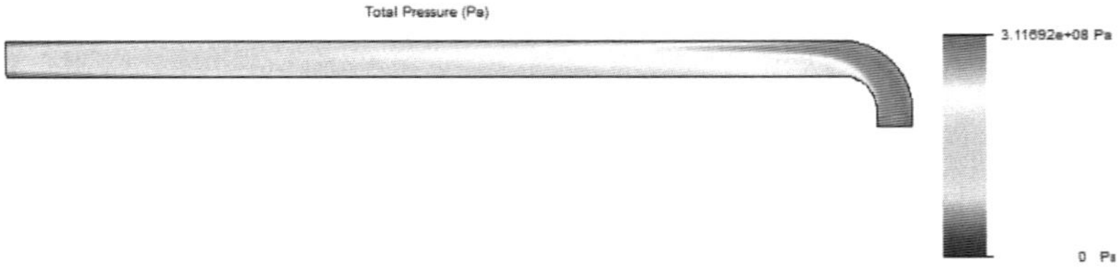

(a)

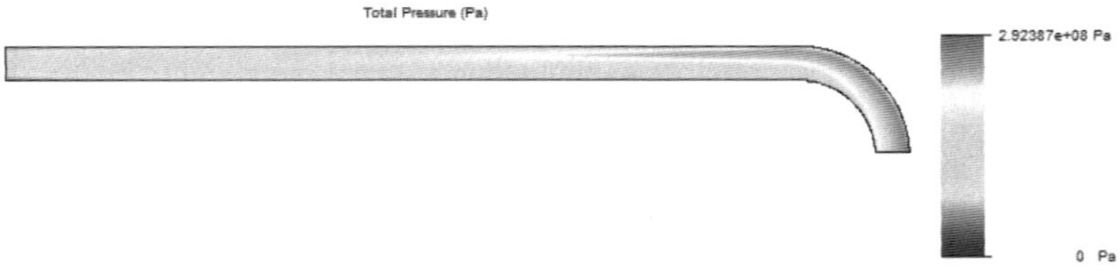

(b)

Figure 7. Maps of the total pressure variation distribution along the studied powered support leg securing system geometries for a fluid volumetric flow of 2000 $dm^3 \cdot min^{-1}$: a – pressure variation distribution for a powered support leg securing system geometry from Figure 2, b – pressure variation distribution for a powered support leg securing system geometry from Figure 4.

By analysing the curves in Figure 5 it can be observed that increasing the leg securing system bend radius by 60% results in an increase of the stream total pressure at the powered support at the outlet of pressure limiting valve by 3 MPa on average. Furthermore, by observing the maps of the total pressure variations along the analysed connector variants, it can be concluded that increasing the connector bend radius decreases the fluid resistances at the inlet by 10% on average for a volumetric flow of 1000 $dm^3 \cdot min^{-1}$ and by 40% on average for a volumetric flow of 2000 $dm^3 \cdot min^{-1}$.

4.2. Stream cross-section influence

In order to analyse the variant involving the influence of the fluid cross-section variation on the pressure distribution along the powered support leg securing system, numerical simulations were conducted for the port assembly model, where the pipe internal diameter (d_{in}) and external diameter (d_{out}) were modified. The basic diameter values, i.e. d_{in} = 20 mm and d_{out} = 30 mm, were increased to d_{in}= 30 mm and d_{out} = 40 mm. Figure 8 presents the basic dimensions of the modified connector (Figure 8a) and the view of the model in the assembly with the pressure limiting valve block (Figure 8b).

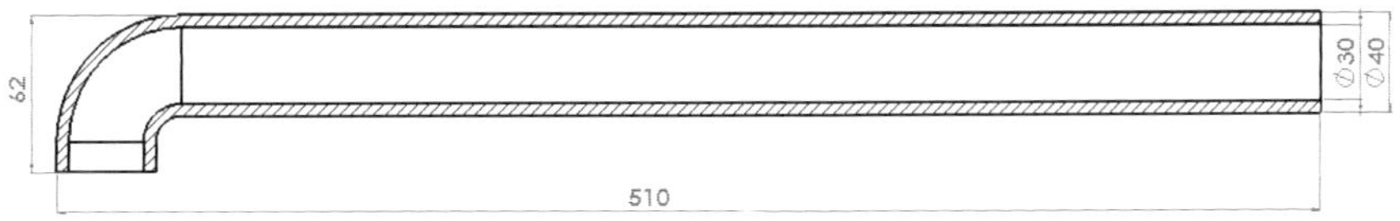

(a)

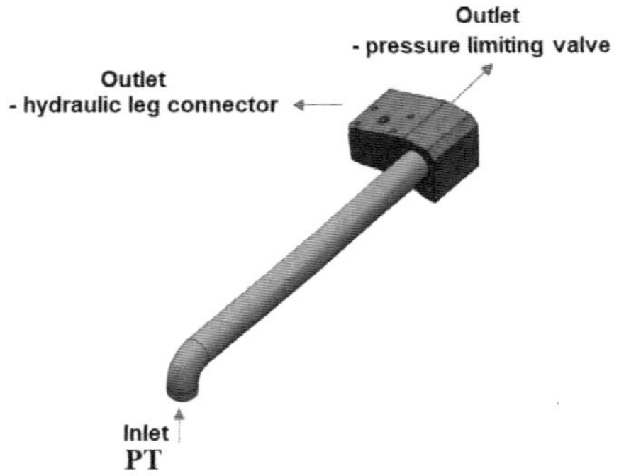

(b)

Figure 8. View of the modified powered support leg securing system geometry: a – connector cross section, b –view of leg securing system

The numerical calculations were conducted within the scope of water volumetric flow variations at a range of 0 to 2000 $dm^3 \cdot min^{-1}$, simulating the fluid supply to the modified and unmodified leg securing system, as depicted in Figure 8. Total pressure variation measurements were performed at the outlet of the limiting pressure valve block during the conducted modelling.

The results of the numerical analysis of the modified port are presented in Figure 9 in the form of powered support leg securing system total pressure variation characteristics.

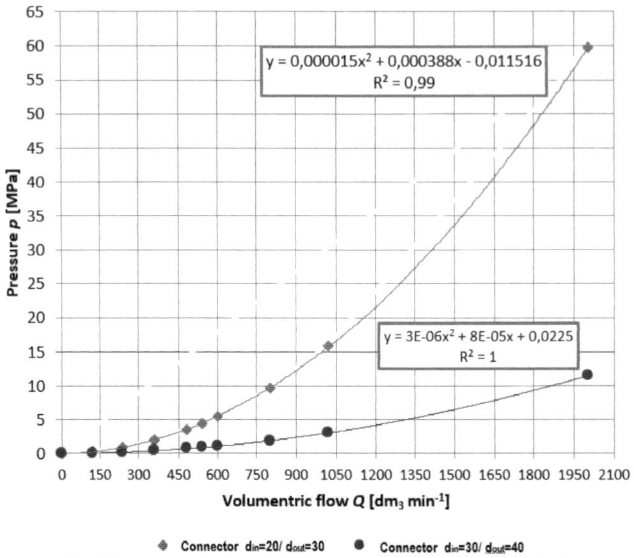

Figure 9. Characteristics of the total pressure variation along the studied geometries at the outlet of leg securing system depending on the fluid volumetric flow.

Additionally, maps of the total pressure variation distribution along the studied powered support leg securing system for a fluid volumetric flow of 1000 $dm^3 \cdot min^{-1}$ are presented in Figure 10 as well as for a fluid volumetric flow of 2000 $dm^3 \cdot min^{-1}$ in Figure 11.

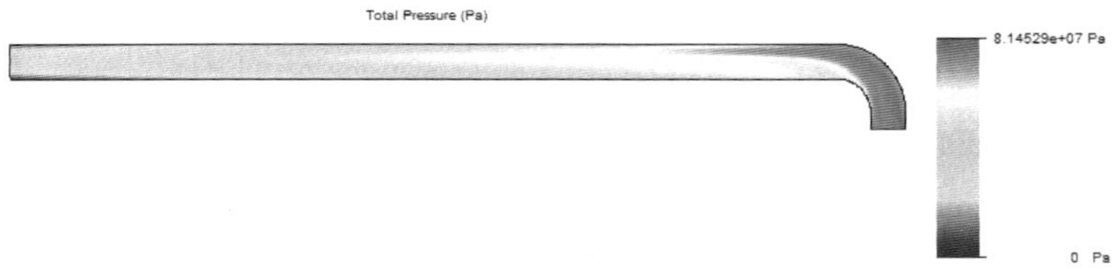

(a)

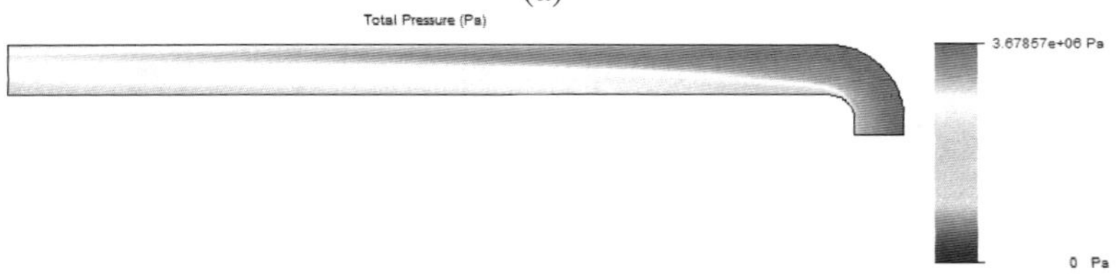

(b)

Figure 10. Maps of the total pressure variation distribution along the studied powered support leg securing system for a fluid stream volume of 1000 $dm^3 \cdot min^{-1}$: a – pressure variation distribution for leg securing system geometry from Figure 2, b – pressure variation distribution for leg securing system geometry from Figure 8.

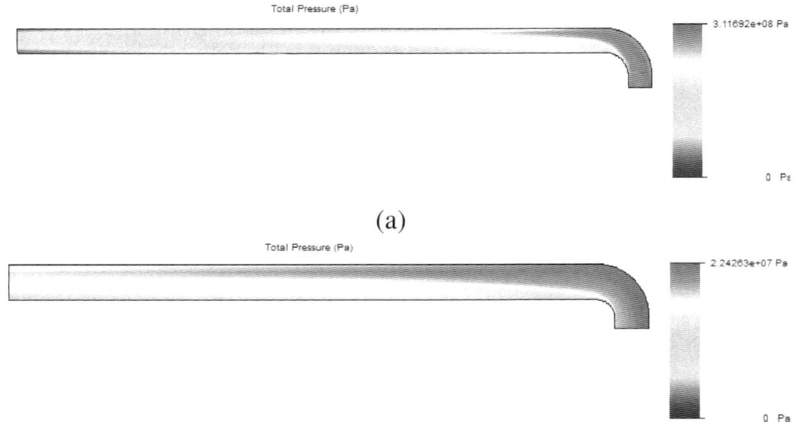

(b)

Figure 11. Maps of the total pressure variation distribution along the studied powered support leg securing system for a fluid volumetric flow of 2000 $dm^3 \cdot min^{-1}$: a – pressure variation distribution for leg securing system geometry from Figure 2, b – pressure variation distribution for leg securing system geometry from Figure 8.

It can be observed from the courses of the curves in the charts (Figure 9) that increasing the cross-section of connector by 25% can result in an increase in the total pressure at the outlet of limiting pressure valve block by 80% on average.

4.3. Stream air content influence

In order to analyse the variant involving the possibility of the occurrence of leaks in the leg securing system assembly, a computational variant of the influence of the air content in the fluid volumetric flow on the total pressure distribution along the leg securing system was considered, measured at the outlet, at the pressure relief valve attachment point. The calculations were conducted for a mixture of water and air with air mass fraction range of 0 to 4%. A mixture was modelled in proportion of air to water: 1:49 and 1:24 . The results are presented in Figure 12 in the form of total pressure variation characteristics at the outlet of the limiting pressure valve block, as a function of the fluid volumetric flow variation and the air content in the fluid (mixture of water and air).

IOP Conf. Series: Materials Science and Engineering **679** (2019) 012017 doi:10.1088/1757-899X/679/1/012017

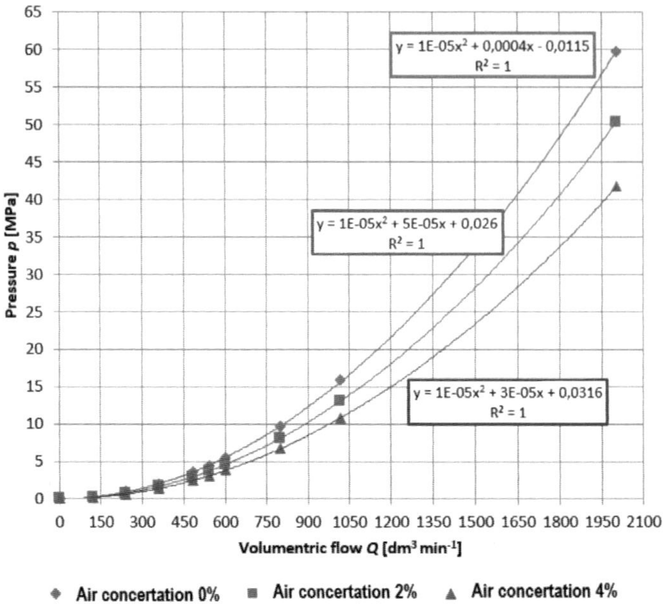

Figure 12. Characteristics of the total pressure variation at the powered support at the outlet of leg securing system depending on the fluid volumetric flow and the air content in the fluid (mixture of water and air).

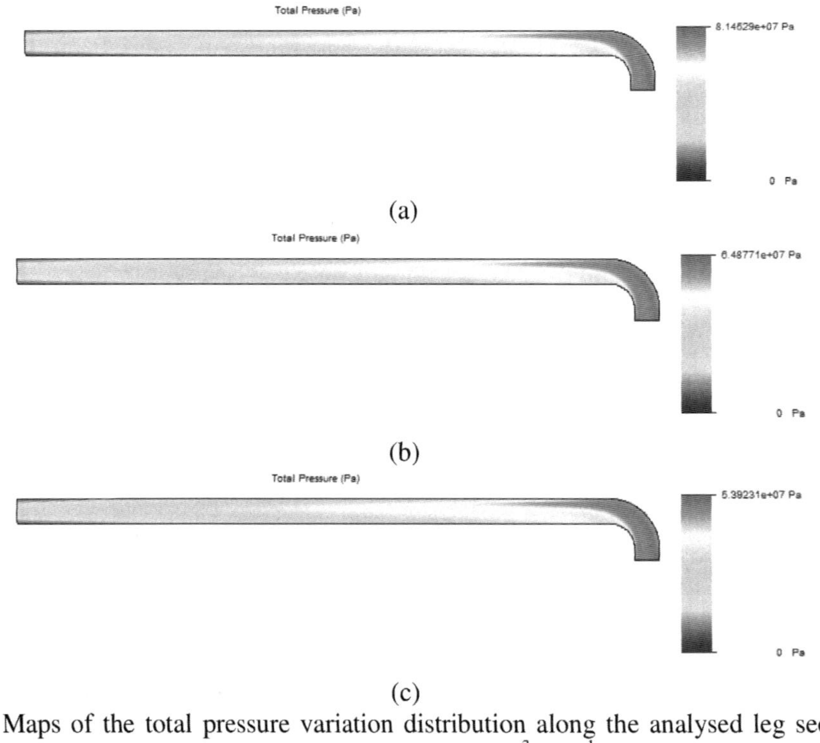

Figure 13. Maps of the total pressure variation distribution along the analysed leg securing system according to Figure 2 for a fluid volumetric flow of 1000 $dm^3 \cdot min^{-1}$: a – pressure variation distribution for an airtight powered support leg securing system assembly, b – pressure variation distribution for a powered support leg securing system with an air mass fraction of 2%, c – pressure variation distribution for a powered support leg securing system with an air mass fraction of 4%.

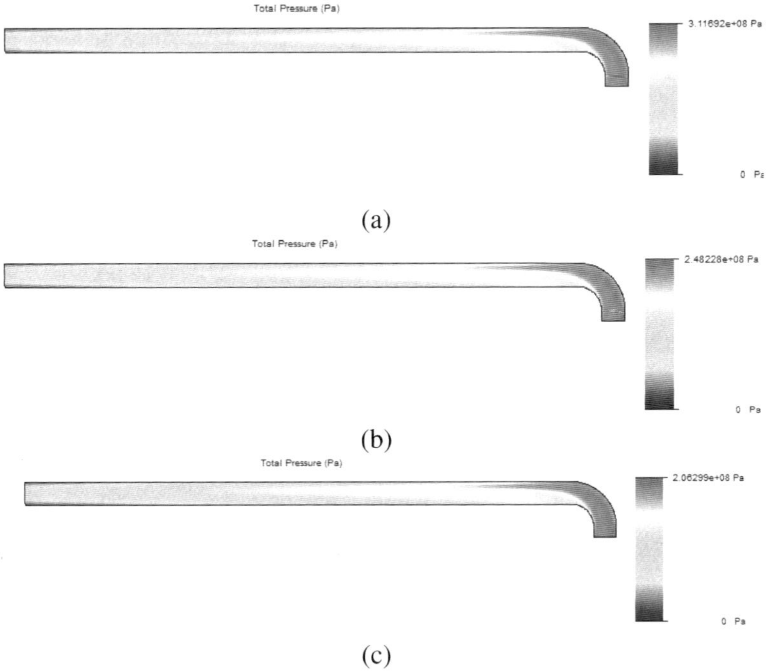

Figure 14. Maps of the total pressure variation distribution along the analysed powered support leg securing system according to Figure 2 for a fluid volumetric flow of 2000 $dm^3 \cdot min^{-1}$: a – pressure variation distribution for an airtight powered support leg securing system assembly, b – pressure variation distribution for a powered support leg securing system with an air mass fraction of 2%, c – pressure variation distribution for a powered support leg securing system with an air mass fraction of 4%.

By analysing the chart of the total pressure variations at the outlet depicted in Figure 12, it can be concluded that an increase in the fluid of air mass fraction within a range of 0 to 4% results in a total fluid (mixture of fluid and air) pressure drop measured at the outlet by 30% on average.

5. System safety characteristics

Flows in the leg hydraulic securing system have a significant influence on the limitation of powered support unit overloads generated as a result of rock mass tremors. The flow values are determined for the entire system encompassing the pressure relief valve and the applied port connecting the space under the leg piston with the valve. The flow value has an influence on vibration damping in the computational model used for the evaluation of the longwall powered support unit yield. The pressure relief system flow in the leg should secure the leg up to a value 1.5 times as great as the base yield load [10, 11]. The flow in the system is presented below in the form of an exponential polynomial for a case of the modification of the internal and outer diameters of the applied pipe. Connector flow determined analytically by means of *SolidWorks Flow Simulation* was considered for two cases of connector diameters. Valve characteristics are presented according to the tests conducted at TLO Opava in Czech Republic [10]. After obtaining the characteristics of the leg securing system and the valve, it was possible to characterise the entire system, as depicted in Figure 15.

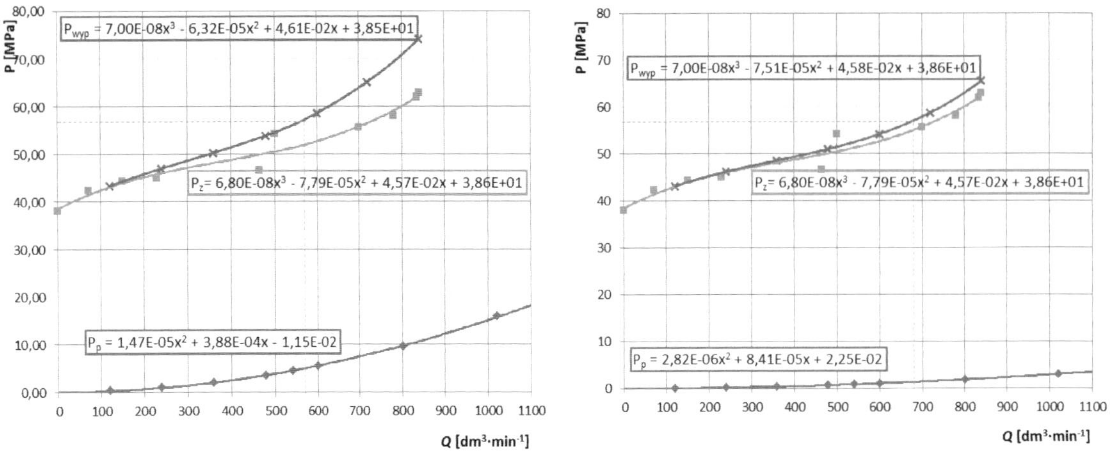

(a) (b)

Figure 15. Flow analysis for the evaluation of the entire system securing an example leg from overloading (P_r = 38 MPa, Q for $1.5 \cdot P_r$ = 57 MPa); P_{wyp} – resultant pressure of the entire system for a defined mass flow, P_z – valve characteristics, P_p – connector characteristics, P – pressure, Q – flow.

As can be seen in the above example, increasing the pipe internal diameter results in an increase in the flow, which is particularly visible at the higher volumetric flow range.

6. Conclusions

The numerical determination of longwall powered support unit leg safety system volumetric flows is useful when evaluating support yield for cases of the extraction of deposits at risk of rock mass tremors, as a way of fulfilling the requirements of the Ordinance of the Minister of Energy of 23 November 2016, Dz. U. No 2017, item 1118, para. 523, sec. 1, pt. 1. They also constitute a significant benefit for designers in terms of the optimisation of the construction of the safety elements that secure the leg against overloads.

The conducted numerical analyses have shown that the fluid stream geometry and its cross-section, resulting from the diameter of the applied pipe, have a significant influence on the limitation of losses in fluid flow from the secured leg space to the pressure relief valve. These parameters attain particular importance at high flow values.

The air content in the analysed stream, up to a value of several percent, had minor influence on the conducted analyses. However, additional calculations should be performed for higher air concentrations in the stream in order to verify the influence of the air content.

It should also be noted that the value of the losses in stream flows is dependent on multiple parameters, the majority of which exhibit non-linear characteristics, which is why it would be the most rational to consider every case individually.

Acknowledgement

The article was prepared based on research conducted within the Research Project: Productivity and Safety of Shield Support (PRASS III)-co-financed by European Commission-Research Fund for Coal and Steel (No. 752504) and Polish Ministry of Science and Higher Education.

References

[1] Janoszek T, Łączny MJ, Stańczyk K, Smoliński A, Wiatowski M 2013 Modelling of gas flow in the underground coal gasification process and its interactions with the rock environment (J. Sustain. Min.) vol. **12**, no. 2, pp. 8–20

[2] Jonuskaite A 2017 Flow simulation with SolidWorks (Bachelor Thesis, Plastics Technology)

[3] Małkowski P 2017 Management of Mining Hazard Monitoring (*J. Polish Miner. Eng. Soc.*) vol. **18**, no. 2, pp. 215–224

[4] Prusek S, Płonka M, Walentek A 2016 Applying the ground reaction curve concept to the assessment of shield support performance in longwall faces (*Arab. J. Geosci.*), vol. **9**, no. 3, pp. 1–15

[5] Prusek S, Rajwa S, Wrana A, Alicja K, 2017 Assessment of roof fall risk in longwall coal mines, (*Int. J. Mining, Reclam. Environ*) vol. **31**, no. 8, pp. 558–574

[6] Rajwa S, Janoszek T, Prusek S 2019 Influence of canopy ratio of powered roof support on longwall working stability – A case study (*International Journal of Mining Science and Technology*), Volume **29**, Issue 4, pp 591-598

[7] Rajwa S, Masny W, Wrana A 2017 A comprehensive method for the selection of powered roof support in conditions of the rockburst hazard *Wiadomości Górnicze* No. **1** pp 2-7

[8] Rajwa S, Prusek S, Stoiński K 2016 Opis metody upodatnienia zmechanizowanej obudowy ścianowej *Bezpieczeństwo Pracy i Ochrona Środowiska w Górnictwie* No **12** pp 3-8

[9] SolidWorks Help SolidWorks Flow Simulation Overview (https://help.solidworks.com/2018/english/SolidWorks/floxpress/c_flow_simulation_overview.htm)

[10] Stoiński K et al. 2018 *Zmechanizowane obudowy ścianowe dla warunków zagrożenia wstrząsami górotworu* Katowice: Główny Instytut Górnictwa, 2018

[11] Stoiński K, Prusek S, Rajwa S 2015 Dobór przepływów w układzie zabezpieczającym stojak zmechanizowanej obudowy ścianowej – opis przypadku (XXIII International Conference TEMAG, Gliwice)

[12] Świątek J Stoiński K 2019 Case Analysis of Damages to Control Hydraulics of the Leg in the Powered Roof Support Section (IVth International Innovative Mining Symposium, E3S Web Conf.) Volume 105

IOP Conf. Series: Materials Science and Engineering **679** (2019) 012018 doi:10.1088/1757-899X/679/1/012018

Comparative analysis of hydraulic power systems for a powered roof support aimed at obtaining optimal power parameters

P Szolc[1], D Szurgacz[1]* and K Styrylski[2]

[1]Polska Grupa Górnicza S.A., Oddział KWK ROW ruch Chwałowice, Przewozowa 4, 44-206 Rybnik, POLAND.
[2]PepKon Firma inżyniersko – informatyczna, Katowice POLAND

*E-mail: dawidszurgacz@vp.pl

Abstract. Hydraulic power systems for powered roof supports, depending on the place of installation of hydraulic aggregates, can be divided into local and central systems. Central systems are gaining popularity and a number of their features are more advanced compared to the local system. However, it is widely believed that their main disadvantage is the necessity to install pipelines with larger lengths, which can have a significant impact on the obtained hydraulic power supply parameters by occurring pressure losses on the pipeline length. This paper discusses the impact of these parameters on the correct operation of the powered roof support section, which is connected with the uninterrupted continuity of the production cycle in a longwall.

1. Introduction

The increase in the global demand for energy raw materials, such as coal, forces underground mining plants to maintain the current level of extraction and even to increase the output [2]. The increase in production may be obtained by operating in deeper seams. This in turn leads to an increase in natural hazards. One of such threats is methane, which significantly limits the progress of extraction [10, 13]. In order to maintain the current state of extraction and to influence its increase, intensive research is carried out on methane hazard modelling in order to effectively combat it [8, 10, 12].

Currently, in the research concerning the hydraulic system of the powered roof support section, much attention is devoted to control systems and active pressure monitoring in the under-piston space of a hydraulic leg [5, 6, 7]. Thanks to the rapid development of electronics, information technology and automation, particular emphasis is placed on the modernization of electrohydraulic control systems, striving to develop appropriate methods and algorithms enabling fully automatic operation of powered roof supports in shearer's mechanization systems [1, 5, 9]. These activities are to be adopted primarily to increase the level of safety and improve the efficiency of extraction. However, for these systems to fulfil their functions, the powered roof support section should be supplied with an emulsion at an appropriate flow rate and under appropriate pressure. These parameters are to guarantee a properly selected hydraulic power system.

The basic elements of hydraulic aggregates used in hydraulic power systems of the powered roof support are emulsion tank, pump units, high pressure and flow filter assemblies. Figure 1 shows a schematic diagram of a hydraulic unit for powering a powered roof support. The idea of operation and construction of hydraulic aggregates in the applied power systems is almost always identical, and the individual elements differ mainly in the parameters and technical solutions used. Emulsion from the reservoir through the suction line (marked in blue) goes to the pumps, from where it is pumped through the filter into the pressure pipe under high pressure. After powering the roof support, the emulsion flows through the discharge pipe to the tank through the filter.

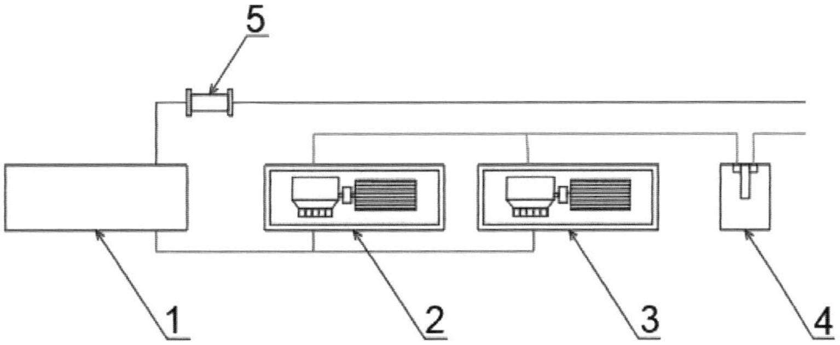

Figure 1. Schematic diagram of a hydraulic aggregate: 1. Oil-and-water emulsion tank, 2. Main pump unit, 3. Additional pump unit, 4. High pressure filter, 5. Outflow filter.

Depending on the place of installation of hydraulic aggregates and the number of powered longwall complexes, hydraulic power systems of the powered roof support can be divided into local and central systems (Figure 2). The local systems are located in the mining region close to the mining wall, most often in the collective slip or in the access drift. Typically, this system powers the powered support sections located within one exploited wall. This is the most commonly used solution in Polish mines so far. The central system is characterized by the fact that hydraulic aggregates are usually located near the shaft in a specially prepared and separated chamber. The central pumping station can power the powered support in several or all of the walls in a given mine. Therefore, the pumping units included in the central pumping station are characterized by higher capacity levels than those usually used.

Due to the location, this central power supply system has a number of advantages over local systems, including:

- obtaining an emulsion with better purity, because the aggregate elements are not in contact with dusty air,
- no need to relocate aggregates and perform work related to the preparation of their location in a way that guarantees correct and safe operation,
- possibility of monitoring pump operation parameters and their diagnostics with the use of automation elements.

However, the central hydraulic power supply system has a significant disadvantage, it is necessary to press the emulsion with pipelines of relatively large lengths. This involves increasing the resistance of the working fluid flow, resulting in a greater reduction in the pressure of the emulsion along the length of the pipeline. Unfortunately, this phenomenon adversely affects forces developed by hydraulic cylinders in the support section.

IOP Conf. Series: Materials Science and Engineering **679** (2019) 012018 doi:10.1088/1757-899X/679/1/012018

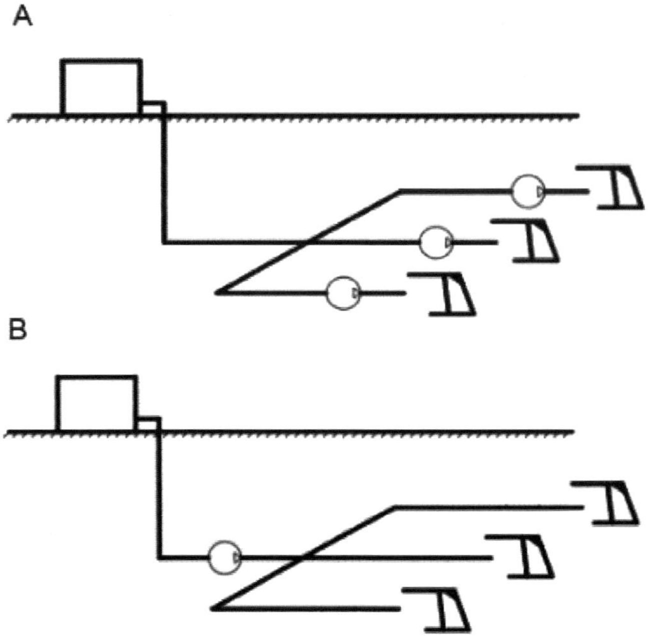

Figure 2. A schematic diagram of hydraulic power systems: A. Local system, B. Central system.

2. Assumptions for powering a roof support

Requirements for the hydraulic power supply systems of the powered roof support are determined by the assumed technical purpose, whose criterion is usually obtaining planned daily excavation. Sections of the powered roof support in the mining wall cooperate with the mining machine and the face conveyor and, thanks to functional and geometrical connections, they form a powered system. This means that the design of the bus supplying the support must be preceded by an analysis of the interaction of individual machines in terms of meeting the assumed technical purpose criterion. As part of such analysis, first of all, the required advance speed of the shearer should be determined to obtain the planned extraction, based on information on geological and mining conditions, technical factors and available technical equipment.

An important parameter affecting the process of completing the hydraulic power supply system is the time in which the support section must be moved to a new location in order to secure the newly developed roof of the excavation [3]. In order to maintain the continuity of the production cycle in the longwall, this time can not be longer than the time the shearer crosses the route equal to the scale of the support section (Figure 3), in other words, the roof development speed should be at least equal to the required advance speed of the shearer.

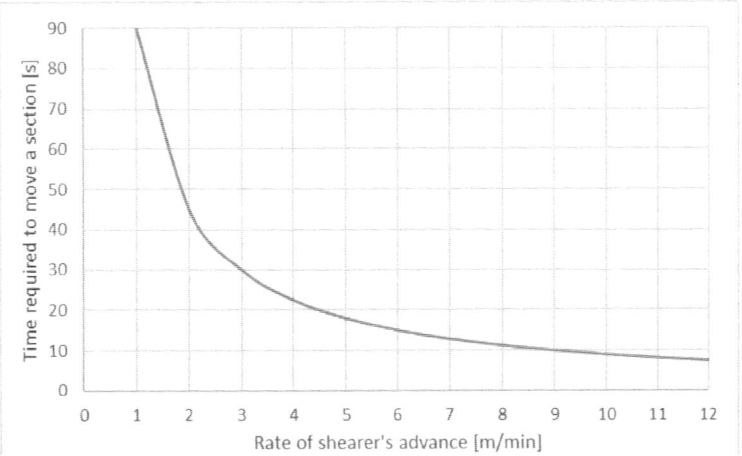

Figure 3. Relationship between the time of advance of the powered roof support and the speed of the shearer [3].

For the assumed assumption, the minimum required emulsion flow rate in the supply system of the support section can be determined. The intensity should be such as to ensure that all activities related to moving the section to a new location within the determined time (Figure 4). The absorptivity of the powered roof support section will be influenced by the parameters of the hydraulic cylinders, the control system, and the number of control functions simultaneously carried out. In the case of high required roof installation speeds, it is worth considering the dynamic properties of the hydraulic system, such as capacitance, which affects the reaction time of the system during calculations. In order to determine the total required capacity of the hydraulic power unit, it is also necessary to pay attention to the loss of emulsion due to leaks and to the absorptivity resulting from moving the face conveyor [3].

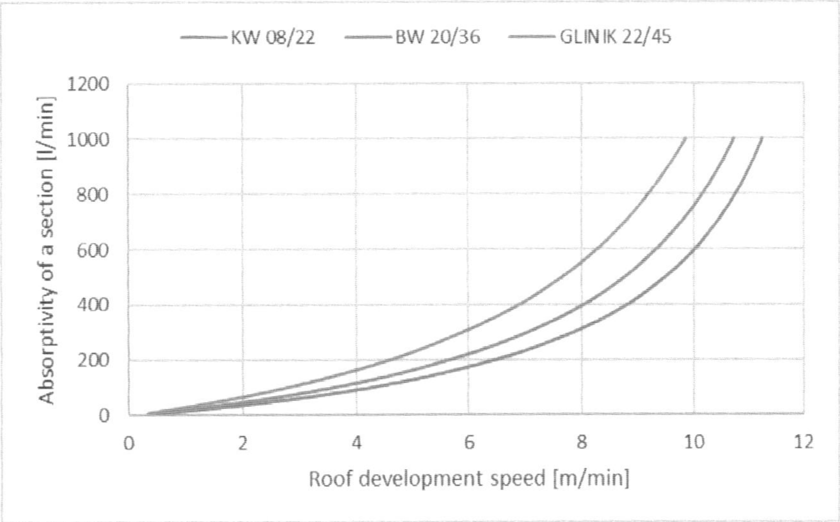

Figure 4. Theoretical absorptivity of selected powered roof support sections depending on the roof development speed [3].

IOP Conf. Series: Materials Science and Engineering **679** (2019) 012018 doi:10.1088/1757-899X/679/1/012018

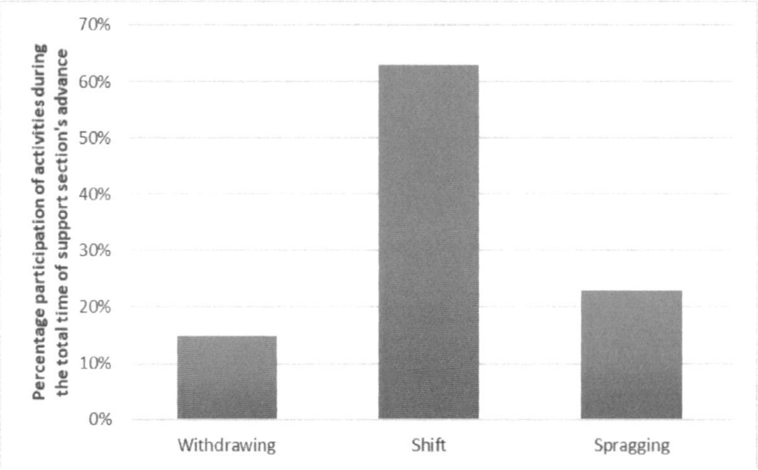

Figure 5. Percentage participation of individual activities in relation to the total time of moving the powered roof support unit KW-08/22-POz/ZRP.

The flow resistance of the working medium and the resulting pressure losses in the supply and control system, in particular the hydraulic shifter, also significantly affect the time of the support section advance [4]. The research team used measurements of the time of the KW 08/22 POz/ZR support sets advance in the mining longwall and theoretical relations to determined that the section moving operation occupies the largest part of the total advance time (Figure 5. Percentage participation of individual activities in relation to the total time of moving the powered roof support unit KW-08/22-POz/ZRPFigure 5) and is the most important in terms of obtaining the required roof development speeds. It can also be noticed that the operation of sliding out of leg lasts longer than sliding down. This results from the fact that it is necessary to recharge the pressure in the pressure bus. The supply bus should be constructed in such a way as to limit the occurrence of pressure losses.

3. Pressure losses in pipelines feeding a longwall system

Optimal selection of pressure and flow pipelines requires knowledge of pressure losses on their length depending on the flow of the working medium. The oil-water emulsions in coal mines are usually supplied through galvanized pipes. Figure 6 shows the pressure loss in the pipeline depending on the flow rate of the working fluid, the analysis covered the 4 most commonly used nominal diameters. The loss of pressure decreases with increasing diameter of the hydraulic conduit. On the basis of the analysis of the obtained results, it is not recommended to use pipes with a nominal diameter of 32 mm, when it is necessary to press emulsions with a high flow rate or a long distance [3].

IOP Conf. Series: Materials Science and Engineering **679** (2019) 012018 doi:10.1088/1757-899X/679/1/012018

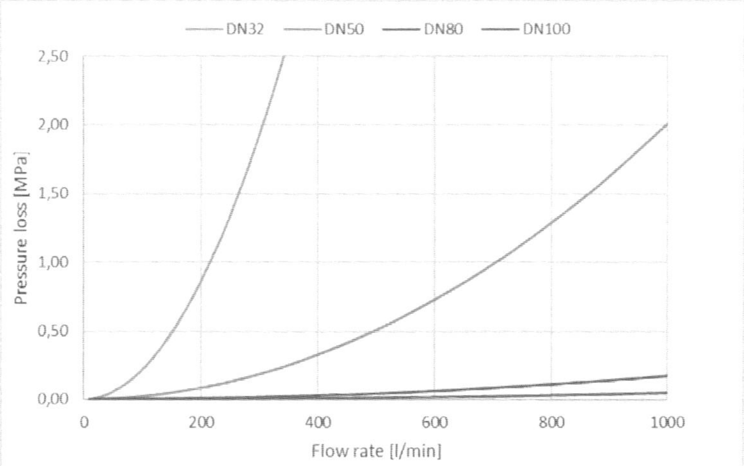

Figure 6. Pressure loss on a section of 100 m pipeline comprising pipes of a given diameter.

4. Comparison of pressure losses in the central and local hydraulic supply systems

The comparative analysis included the case of supplying one of the excavations of KWK ROW Ruch Chwałowice mine. It is a longwall III in part IIIz of the mine in seam 408/1, which was decommissioned as the resources were depleted. The powered roof support located in this longwall (KW-08/22-POz) was powered by the local system during normal operation. Obtaining planned daily output was possible thanks to the use of 4 pump sets type T-100/32 manufactured in Poland (Figure 7). The total performance obtained in this way could amount to 400 l/min, which corresponds to the efficiency achieved by one Inoxihp PM400 pump (Figure 8) used in a central hydraulic pumping station.

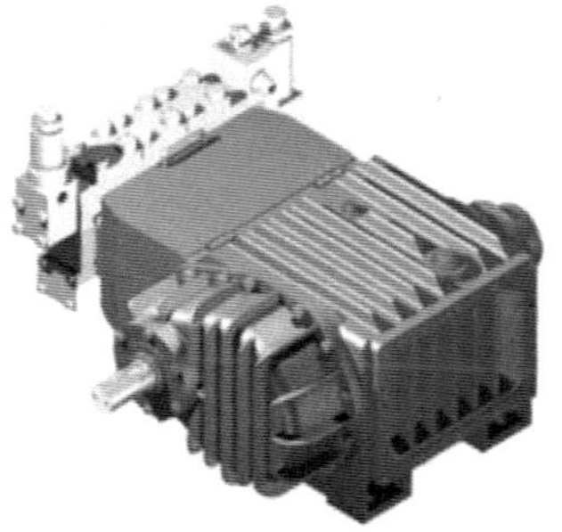

Figure 7. Pump type T-100/32.

Figure 8. Pump type Inoxihp PM400.

Figure 9 presents a schematic diagram of the underground infrastructure of the KWK ROW Ruch Chwałowice mine. It was prepared on the basis of mining maps that enabled obtaining information on the depth of the given excavations, their length and mutual location. Thanks to the knowledge of the difference in the depths of individual workings, static pressure in the pipeline can be determined, which is particularly important from the point of view of the flow pipeline. The length of the excavations, which corresponds to the length of the pipelines installed in them, affects the resistance of the working fluid flow. The analysis presented in this paper covered only the length of excavations, because in the case under consideration the differences in depth did not play a significant role in comparing the two hydraulic power systems.

The Figure presents the location of the pressure and flow pipeline in the district supply system was marked in a blue, the location of the pipelines in the central system was marked in red. The local power system consisted of:

- DN50 pressure pipeline, 1044 m long,
- DN80 pressure pipeline, 1044 m long.

Pipelines in the central system should be divided into local and header ones. The header pipelines are the main lines where the emulsion goes to the mining regions (they are characterized by a larger diameter), from where individual longwall excavations may be directed by local pipelines. The purposefulness of such a division can be justified by the fact that assembly and disassembly of pipelines with larger diameters is more labour-intensive and time-consuming. Here, the central power supply system consists of:

- DN80 pressure header pipeline, 1228 m long,
- DN100 discharge header pipeline, 1228 m long,
- DN50 pressure local pipeline, 1024 m long,
- DN80 discharge local pipeline, 1024 m long.

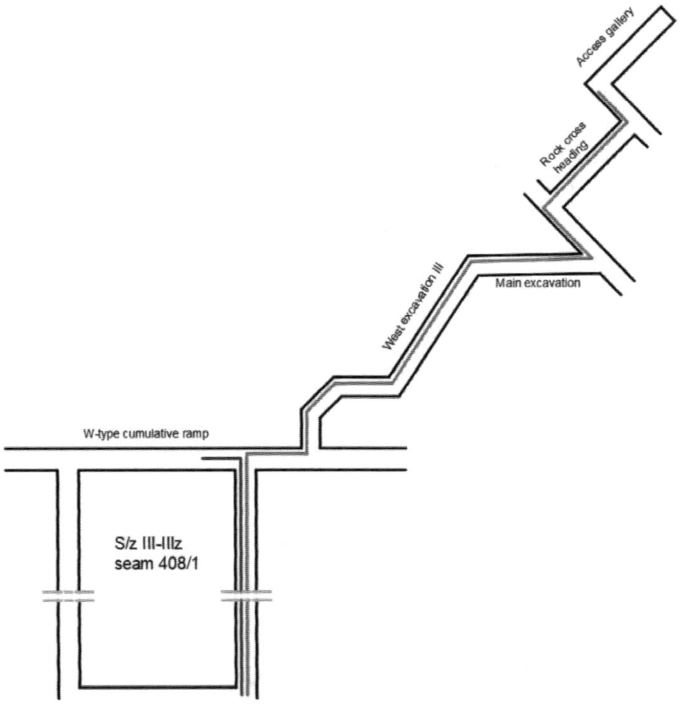

Figure 9. Scheme of underground infrastructure level 700, Chwałowice hard coal mine.

The length of pipelines used in the case of supplying the discussed longwall system is more than twice as high using the central power supply system. In the present situation, this longwall was the only one maintained in the mentioned part IIIz of the mine. Pressure losses were determined in pipelines for a flow rate of 400 l/min. For pipe diameters covered by the analysis, pressure losses per 100 m of pipelines are presented in Table 1. The main reason for the pressure loss in long pipelines is the fluid friction against the conduit walls, local obstacles such as elbows or narrowings have a much smaller effect on pressure loss and are taken into account by increasing linear losses by 10%. Linear losses depend on flow conditions, fluid properties and hydraulic hose parameters. These properties can be described by Reynolds number [14].

Table 1. Pressure loss over a 100 m section of the pipeline with a given diameter at a flow rate of 400 l/min.

Pipe's diameter (mm)	Pressure loss (MPa)
32	3.4
50	0.33
80	0.029
100	0.009

Based on the results presented in Table 1, pressure losses in the discussed systems were determined together with 10% of local losses:

- pressure pipeline in the local system – 3.79 MPa,
- discharge pipeline in the local system – 0.33 MPa,
- pressure pipeline in the central system – 4.18 MPa,
- discharge pipeline in the central system – 0.44 MPa,

The difference in pressure losses in individual systems is small and amounts to 0.39 MPa for pressure pipelines, and 0.11 MPa in the case of discharge pipelines. It results from the fact that pipes with larger diameters are installed in header pipelines, thanks to which lower flow velocities and linear loss coefficients were obtained. The flow of oil-water emulsion in the pipes of the hydraulic supply systems of the powered roof support is most often turbulent, which means that the fluid particles move in a disordered manner. The pressure loss along the length of the hydraulic hose is proportional to the square of the speed, therefore it is very important to obtain low flow rates of the working medium. An increase in the diameter of the hydraulic hose results in a decrease in the Reynolds number, which results in a more favourable linear loss factor [14]. In this case, replacing the local system with the central one did not significantly worsen the hydraulic power supply conditions. However, this does not change the fact that the pressure loss in the supply pipe alone was 13.9% of the supply pressure. The analysis also included calculations for the case of a central system in which the flow rate in the header pipelines is 800 l/min and in the local 400 l/min. This assumption is to simulate the supply of two longwall systems. In this case, the difference between the pressure loss in the local and central system was slightly higher and amounted to 1.35 MPa, the total pressure loss was 17.3% of the supply pressure. The distribution of losses in pipelines is presented in Figure 10 and Figure 11.

In both cases, however, the pressure pipeline DN50 installed in the bottom gate, not the type of hydraulic power system used, has a decisive influence on the pressure loss.

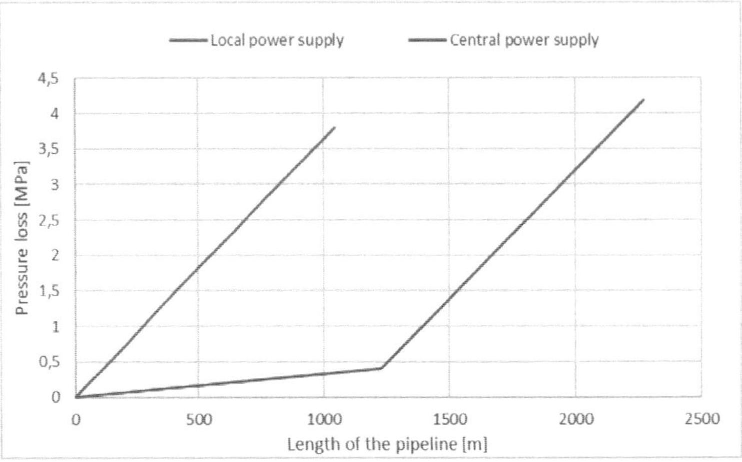

Figure 10. Pressure losses on the length of pipelines in individual hydraulic power systems when the flow rate is 400 l/min.

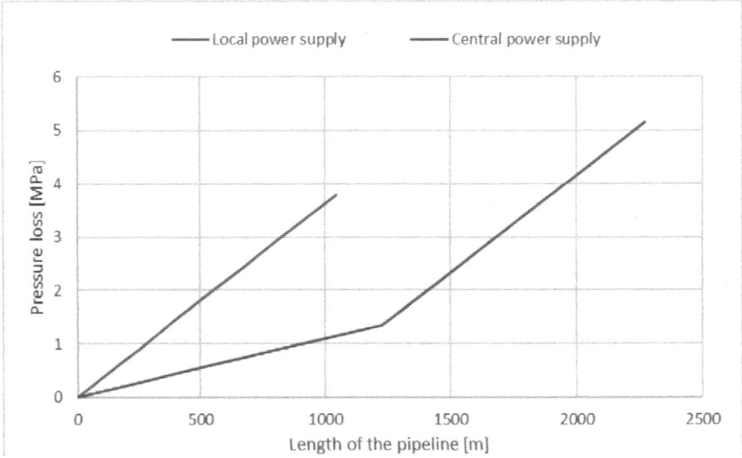

Figure 11. Pressure losses on the length of pipelines in individual hydraulic power systems when the flow rate in the header pipelines is 800 l/min and in the local 400 l/min.

5. Conclusion

The central hydraulic power supply system in mines where the operation is based on longwall system has many advantages over the local system, including [3]:

- better cleanliness of the working fluid,
- a uniform emulsion quality,
- simpler device control - grouped within one chamber,
- possibility of automatic diagnostics, introduction of computers to the central pumping station,
- no possibility of damage to the pumps due to the transport of other materials (it happened despite the maintenance of the correct dimensions of the excavation, that the materials transported to the longwall broke the wires in the pumps),
- no need to relocate devices and prepare for them the right place in the mining area,

- constant supervision of the operator over the operation of pumps,
- short distance to the shaft shortens the transport of the elements required to repair any failures.

The disadvantage of the central hydraulic system of the powered roof support is the necessity of using pipelines of relatively large lengths. The assembly of header pipelines is labour-intensive and time-consuming. It is also necessary to incur higher investment costs. The length of the pipelines also influences the increase of flow resistance of the oil-water emulsion, which is associated with greater pressure losses, which, however, can be eliminated by appropriate selection of pipe diameters.

Thanks to the use of larger diameter pipes, lower flow velocities are obtained which is extremely important because the pressure loss is increasing proportionally in the square of this velocity. The difference in pressure loss in the considered systems was small and amounted to 0.39 MPa in the pressure pipeline and 0.11 MPa in the discharge pipeline, this was obtained by using pipelines with the correct diameter. In the second discussed case, in which the supply of 2 longwalls from the central pipeline of the central hydraulic supply system and one longwall from the local system was simulated, the difference in pressure loss was 1.35 MPa in the pressure pipeline. These differences should not cause significant changes in hydraulic supply conditions between the systems used.

Designing hydraulic power systems for powered roof support systems should be preceded by an analysis of geological and mining conditions, technical factors and technical equipment, so that the selected supply system is an integral part of the longwall system.

Acknowledgements
The work is the result of research conducted as part of the project carried out by Polska Grupa Górnicza S.A. "Learning Organization". Its aim is to strive to actively guide the future of the company.

References
[1] Brodny J and Tutak M 2019 *Analysing the Utilisation Effectiveness of Mining Machines Using Independent Data Acquisition Systems: A Case Study*, Energies 12(13)
[2] Cehlár M, Janočko J, Šimková Z, Pavlik T, Tyulenev M, Zhironkin S, Gasanov M, 2019 *Mine Sited after Mine Activity: The Brownfields Methodology and Kuzbas Coal Mining Case, Resources,* 8(1)
[3] Szolc P 2019 *Określenie założeń i koncepcji centralnego układu zasilania hydraulicznego w zakładzie górniczym PGG S.A. KWK ROW Ruch Chwałowice*. Praca magisterska wykonana w Katedrze Mechanizacji i Robotyzacji Górnictwa Politechniki Śląskiej, Gliwice (not published)
[4] Szurgacz D 2015 Electrohydraulic control systems for powered roof supports in hazardous conditions of mining tremors, *Journal of Sustainable Mining* Volume **14**. Issue 4, doi.org/10.1016/j.jsm.2015.12.001
[5] Szurgacz D and Brodny J 2018 *Role and relevance of an electro control systems for the operation of a powered roof support,* 17th International multidisciplinary scientific geoconference SGEM, 29 June - 5 July, Albena, Bulgaria. Conference proceedings Volume 17. Science and technologies in geology, Exploration and Mining Issue 13. p. 781-788, doi.org/10.5593/sgem2017/13/S03.099
[6] Szurgacz D and Gondek H 2018 *Test of an innovative controller designed to control the powered roof support*, XV Internationa Conference Multidisciplinary Aspects of Production Engineering 05-08 september 2018 Zawiercie, Poland. Conference proceedings Volume 1, Issue 1, s. 223 -231, 2018 doi:10.2478/mape-2018-0029
[7] Szurgacz D and Brodny J 2018 *Analysis of rock mass dynamic impact influence on the operation of a powered roof support control system*, E3S Web of Conferences 29, 00006 XVIIth Conferences of PhD Students and Young Scientist doi.org/10.1051/e3sconf/20182900006

[8] Szurgacz D, Sobik L, and Brodny J 2019 *Ventilation systems in longwall workings with powered longwall complex. Geolinks International Conference on Geosciences.*,Conference proceedings, Book 1, Volume 1. p. 201 – 209. 26-29 March, Novotel, Athens, Greece

[9] Szurgacz D, Szolc P and Cuber G 2019 Udoskonalanie kompleksu ścianowego w oparciu doświadczenia ruchowe na przykładzie KWK ROW Ruch Chwałowice. Systemy Wspomagania w Inżynieri Produkcji, Górnictwo – Perspektywy, Zagrożenia Vol.**8** issu 1, s. 439-444

[10] Tutak M and Brodny J 2017 *Analysis of Influence of Goaf Sealing from Tailgate On the Methane Concentration at the Outlet from the Longwall*, IOP Conf. Series: Earth and Environmental Science, 95, 042025 doi:10.1088/1755-1315/95/4/042025

[11] Tutak M and Brodny J 2016 *Analysis of gases emitted into the atmosphere during an endogenous fire. Energy and Clean Technologies,* SGEM2016 Conference Proceedings Volumne VIII, ISBN 978-619-7105-82-7, ISSN 1314-2704, Book 4 Vol. 3, pp 75-82. DOI: 10.5593/SGEM2016/HB43/S06.011

[12] Tutak T and Brodny J 2019 *Predicting Methane Concentration in Longwall Regions Using Artificial Neural Networks*, *Int. J. Environ. Res. Public Health,* 16(8)

[13] Tutak M and Brodny J 2018 *Analysis of the Impact of Auxiliary Ventilation Equipment on the Distribution and Concentration of Methane in the Tailgate*, *Energies* 11(11)

[14] Gryboś R 1991 *Mechanika płynów.* Skrypty Politechniki Śląskiej, Gliwice

IOP Conf. Series: Materials Science and Engineering **679** (2019) 012019 doi:10.1088/1757-899X/679/1/012019

Intelligent spraying installation for dust control in mine workings

D Bałaga

KOMAG Institute of Mining Technology, Division of Machinery and Equipment, Pszczyńska 37, 44-100 Gliwice, Poland

dbalaga@komag.eu

Abstract. Design solution of an intelligent spraying installation for dust control in mine workings, developed within the ROCD project is presented. Current solution for reduction of airborne dust in the underground mining industry is described. Design as well as principles of operation of new solution based on the adaptive operation, consisting in adaptation of spraying intensity to the measured dust concentration are discussed. In conclusions, the directions of further research project are pointed out.

1. Introduction

Current requirements regarding the application of dust control devices in hard coal mining industry are focused on the source of dust generation, i.e. when mining coal with the mining machines, during crushing and transferring the run-of-mine on the conveyor belts. Unfortunately, the local method of dust reduction does not eliminate completely the generated dust, contributing to its movement with the air stream of the mine, along the mine workings. This causes that the miners are exposed to the negative effect of dust [1, 2]. The presence of dust in mines, in addition to the explosion hazard, is also the cause of the development of pneumoconiosis, which is the most common occupational disease of miners in hard coal mines. Pneumoconiosis is manifested by chronic bronchitis and emphysema, and sometimes heart failure and hypertrophy, and the effects of dust on the workers' organ are visible only after a few or several years of work (Figure 1) [3].

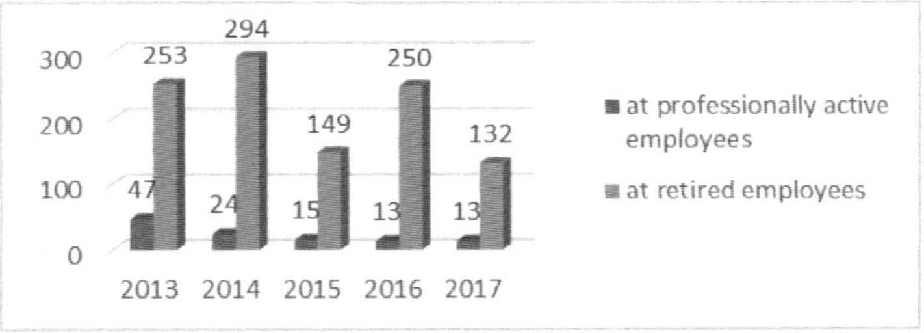

Figure 1. Number of cases of pneumoconiosis observed in the years 2013-2017 among the employees of active coal mines [3].

The suggested concept of an intelligent air-water device intended to prevent against propagation of fine dust particles, i.e. PM2.5 and PM10, to be used in potentially explosive atmospheres will be the response to the demand for dust reduction in roadways, where dust it is carried with air stream. Dust precipitation with use of water drops and its deposition on the floor will be the main task of the installation. The idea of effective dust elimination by the developed device will be based on the generation of a spraying stream of water drops with the diameter close to the size of dust particles, while maintaining a high ejection energy from the nozzle. The solution for producing water drops with diameters close to PM10 and PM2.5, will use the compressed air. Production of spraying streams with small water droplets is to increase the efficiency of dust particles elimination by combining them together, leading to their precipitation. The neutralization process (Figure 2) can take place as a result of combining several transportation mechanisms, the main ones are: inertial, hook and diffusion mechanisms, and in the case of electric potential differences – electrostatic mechanism [4].

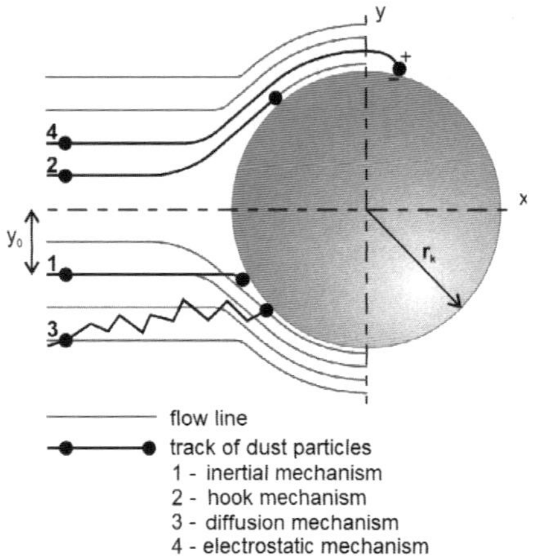

Figure 2. Mechanisms of transporting the solid particles form gaseous phase to the water drop [4].

The mechanism of inertial deposition is based on the impact of a solid particle on the surface of a water drop, the hook mechanism - on its tangential contact, the diffusion mechanism - on the collision of a solid particle with a water drop in a result of random movements, regardless of the stream line, and the electrostatic mechanism - on attracting the opposite electrically charged particles of water and dust.

The spraying installation will be equipped with an intelligent control system which, basing on the information received from the dust sensor about actual dust concentration, will select an adequate intensity of the spraying stream, enabling the reduction of dust concentration below maximum allowable .

2. Review of currently used solutions

Operation of currently used dust control installation, especially in hard coal mines, installed in roadways is based on suction dust from these places [5] or using the water spraying devices [4].

The wet dust control devices are most commonly used to clean airstream from dust particles. Their use allows not only to capture dust particles, but also to neutralize their explosive properties. These devices use the spraying systems in which water is atomized in the air stream and the dust particles penetrate into the water droplets. Regardless of how the particles are mixed with water drops, a mixture of air with water drops and dust particles is swirled, while the drops, due to the centrifugal force, are directed towards the walls of the dust collector, where ribbed surfaces stop them and they flow to the water reservoir. The water circuit in the dust control device is schematically shown in Figure 3.

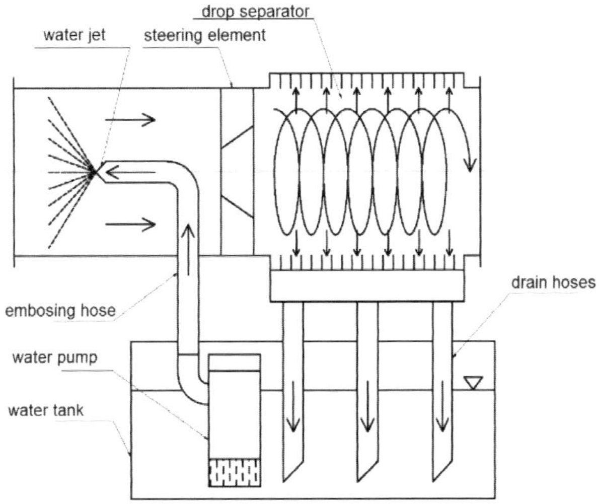

Figure 3. Schematic diagram and principles of operation of the wet dust collector [5].

Dust control by use of water spraying system is one of the most common and cheapest methods used in mines. The sprinkling technique consists in creating a barrier from water mist throughout the cross-section of the excavation, so that the dust is connected with the current of air, it merges with sprayed drops of water, creating moistened agglomerations of dust particles devoid of volatile properties. There are many concepts and design solutions for spraying devices, both domestic and foreign, which are used to reduce dust in pavement excavations. One of the first solutions to reduce dust with water and air, was developed in the 1960s. This solution was intended for use in Soviet hard coal mines (Figures 4 and 5) [6].

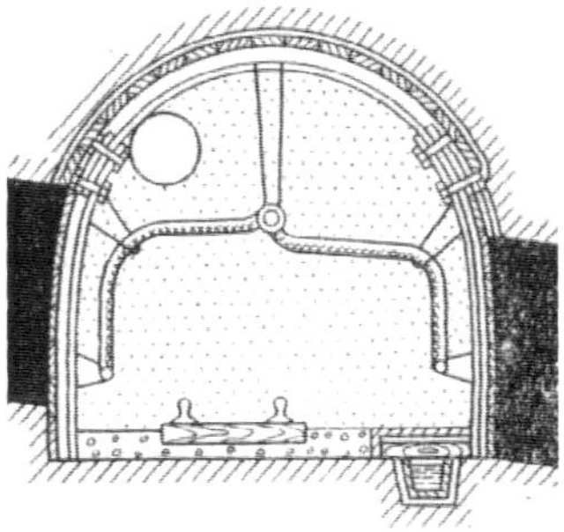

Figure 4. Roadway cross-section with the spraying system [6].

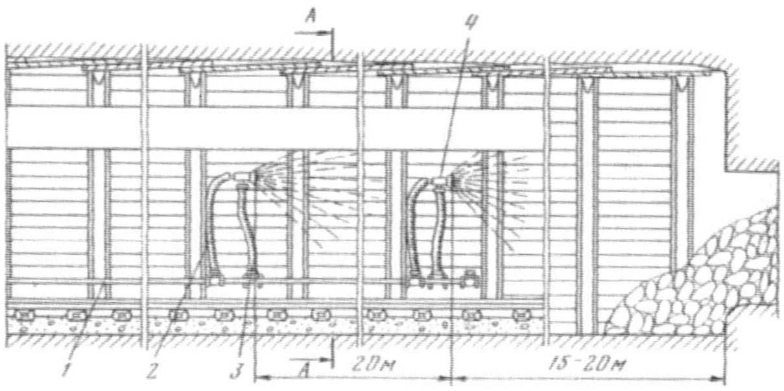

Figure 5. View of a roadway with the installed spraying installation [6].

The system consisted of several nozzles, to which air and water were supplied by means of ducts and was located on a roadway support. This installation was supplied with water and air coming from the pipelines available in the roadway, and the direction of the nozzles was opposite to the direction of the roadway ventilation. Unfortunately, this solution has never been used in underground conditions.

In 2010, a solution of the "mist system with net diaphragms", using air-water spraying system to reduce dust concentration in the roadway, developed by the Telesto Sp. z o.o., appeared in the Polish mines. The solution is also based on the use of water and compressed air to capture dust. The device is made of a spraying installation suspended in the roadway cross-section as well as of moving grids catching water drops combined with dust, in the form of a labyrinth, placed in area of the personnel movement [7]. The spraying system consists of air-water nozzles, located on a frame consisting of pipes supplying water and air to them (Figure 6).

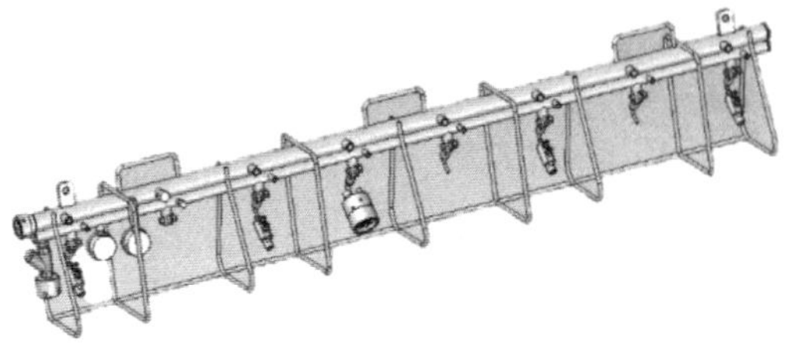

Figure 6. Model of the spraying installation being a part of the mist system [7].

High demand for compressed air is the solution disadvantage. This installation, with a flow of 9 dm^3/min of water and a pressure of 0.6 MPa, needs about 5 m^3/min of air, which is unobtainable in many mines and the need for such air volume raises many problems.

The solution developed in ITG KOMAG called CZP BRYZA (Figure 7) and FOG, where they were equipped with innovative mesh lattices that allow air flow while capturing wetted dust particles (Figure 8) [4, 8] was the answer to high consumption of compressed air in the spraying system. In turn, the air-water nozzles used there enable to achieve high efficiency of dust reduction, with low pressure of spraying media of 0.3-0.5 MPa and low water consumption of 3.0 dm^3/min and compressed air of 0.75÷1.5 m^3/min. This property results from the application of special solutions for spray nozzles in which, already inside, atomization of water takes place due to the use of much smaller amount of compressed air than in the case of a known, external atomization process of water, ultimately providing several times lower air flow.

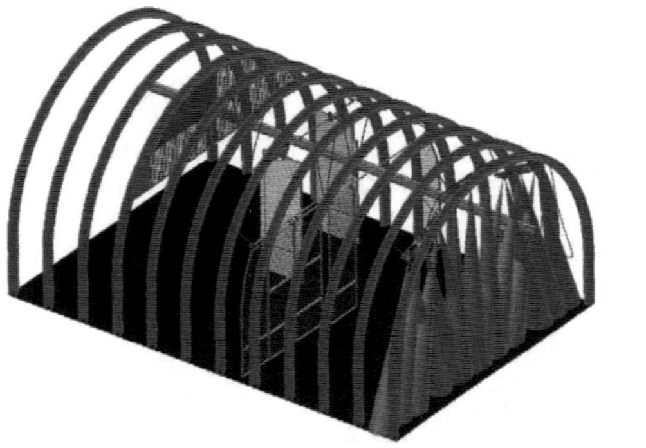

Figure 7. Model of the air-water roadway spraying device CZP BRYZA [4].

Figure 8. Innovative mesh lattices that allow air flow while capturing wetted dust particles [8].

Air-water roadway spraying devices are solutions that significantly contribute to the improvement of safety and comfort of work and they improve the environmental working conditions in roadway workings. They cause a significant reduction in the risk of coal dust explosion. Low efficiency in reducing the PM10 dust, which is the most dangerous fraction for the human body is their important limitation. Underground tests of airborne dust concentration carried out within the project, in the case of using the currently used spraying devices, showed an effectiveness not exceeding 33% (Figure 9).

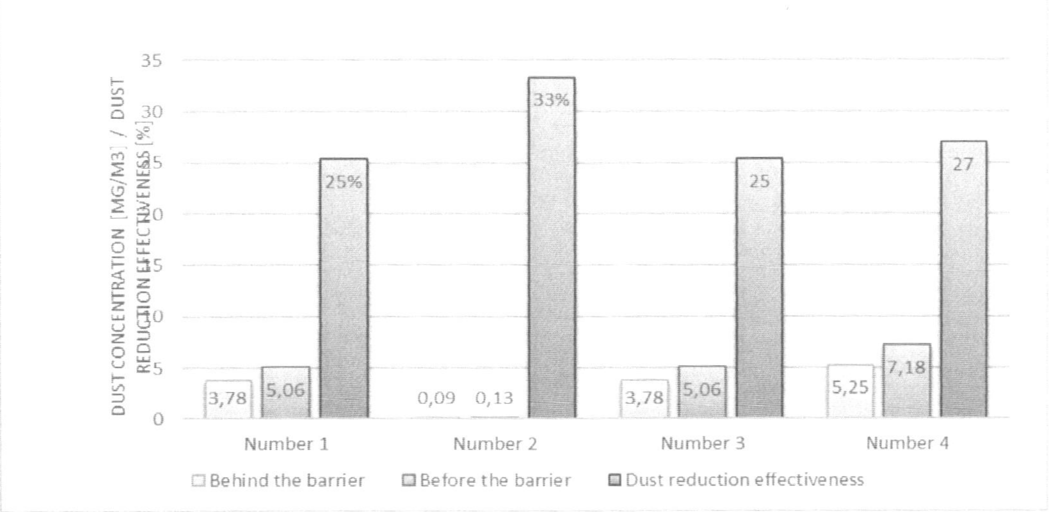

Figure 9. Respirable dust reduction efficiency measured on 4 roadways equipped with currently used spraying devices.

These solutions often cause problems with free passage of the personnel and are characterized by high water consumption resulting from their continuous work, which increase the cost of using such solutions and adversely affects the work comfort of miners. The newly developed concept of an intelligent sparaying installation to be used in roadways, whose main task is to reduce airborne dust concentration is the answer to the above limitations. Such installation should ensure low water and air consumption and high efficiency of capturing even the smallest dust fractions.

3. Concept of intelligent spraying installation

The solution of intelligent spraying installation for controlling airborne dust concentration in mine underground workings was developed within the ROCD project. Adaptive spraying system consisting in adaptation of its output to the dust concentration measured by the EMIDUST dust meter is the solution advantage. The developed spraying installation will use water and compressed air to generate water drops of the size close to the size of PM10 dust.

The intelligent spraying installation (Figure 10) will have dozen or so spraying units installed on the roadway support circumference and they will be fed from the water and compressed air filtration-and-distribution units. Operation of the spraying units will be controlled by the controller, which will decide about spraying duration and intensity depending on PM10 and PM2.5 dust particles concentration measured by a dust meter to reduce airborne dust concentration below MAC.

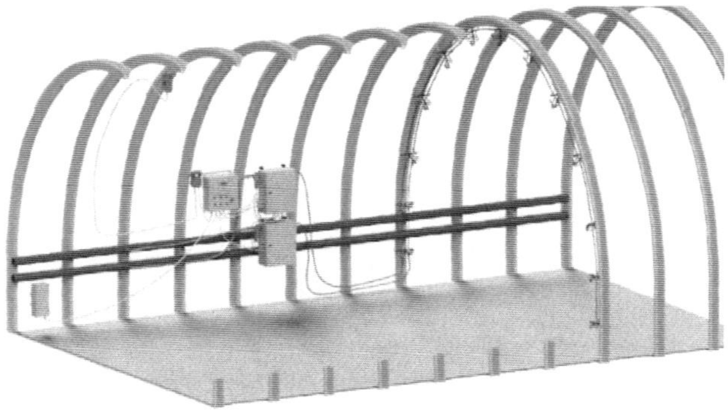

Figure 10. 3D Model of the intelligent spraying installation.

IOP Conf. Series: Materials Science and Engineering **679** (2019) 012019 doi:10.1088/1757-899X/679/1/012019

Water and compressed air will be fed to the water (Figure 11) and compressed air (Figure 12) preparation units, where the media will purified and directed to the spraying units under the required pressure.

Figure 11. Water preparation and distribution unit.

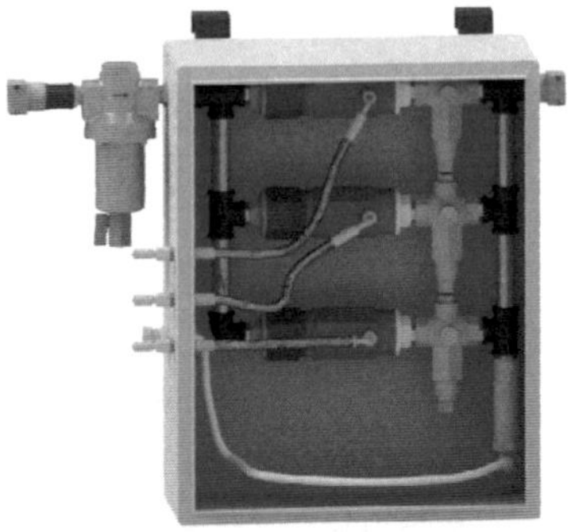

Figure 12. Compressed air preparation and distribution unit.

Water in the water preparation and distribution unit will be divided into three stabilizing-and-supplying lines in which water pressure will be reduced to the set values (three different values Vr_1=0.1MPa, Vr_2=0.2MPa and Vr_3=0.3MPa). Water will flow through each spraying line after opening one of three controlled check valves, directing water stream to the hose supplying the spraying units. By analogy the compressed air in the preparation and distribution unit will be divided into three supplying lines in which air pressure will be reduced by the check valves to the set values (initial settings Vr_1=0.2MPa, Vr_2=0.3MPa and Vr_3=0.4MPa), and then air will be fed to the hoses supplying the spraying units through the controlled check valves (ZZS). Pressure in water and compressed air lines will be controlled by the pressure transducer.

The control of each ZZS controlled check valve is realized by double pre-control valves of Marco GmbH, installed on the distribution board (Figure 13) and it takes place in pairs (activation of one water ZZS check valve and activation of one compressed air ZZS check valve at the same time), so that water and compressed air are supplied to the spraying units.

Figure 13. Double pre-control valves connected to the distribution board.

Electric equipment (Figure 14) of the intelligent spraying installation is responsible for proper operation of the spraying system. Intrinsically safe MDJ controller, responsible for collecting information from the EMIDUST EMAG optical dust meter and from the pressure transmitters of water and compressed air and then deciding about activation of double pre-control valves responsible for water flow to the spraying units, is the main subassembly of the electric equipment. The controller is powered from the ZIS intrinsically safe power feeder.

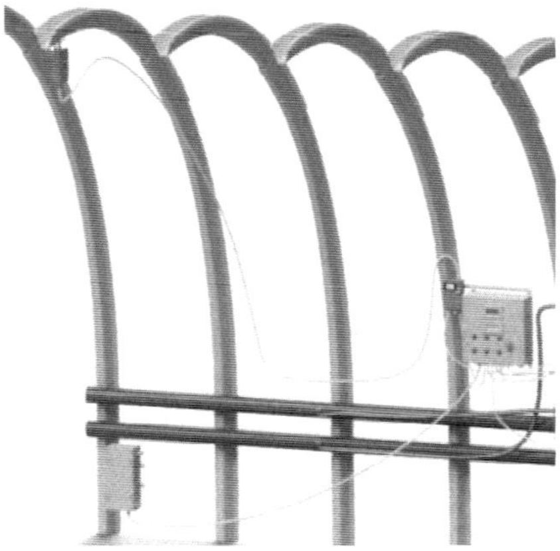

Figure 14. Electric equipment of the intelligent spraying installation.

The intrinsically safe MDJ controller used in the electrical equipment consists of a steel enclosure in which the controller with a display and control buttons was installed. All modules, apart of the central module, are placed in standard plastic enclosures and mounted on a DIN rail fixed to the assembly board. The central module is equipped with a liquid crystal display installed on the enclosure door. Dust meter for the continuous measurement of the PM10 and PM2.5 dust concentrations is another component included in the electric equipment. The solution was developed as part of the project by ITI EMAG and is designed to provide information about the instantaneous concentration of dust sent to the intrinsically safe controller in the protected roadway. The dust meter uses the phenomenon of light scattering on dust particles. The measurement result is presented on the display and in the form of an analog voltage signal.

The spraying unit, generating the stream of water and compressed air is the last component of the intelligent spraying installation (Figure 15).

Figure 15. Spraying unit.

The spraying unit consists of a two-media nozzle, to which the connections of spraying media, i.e. water and compressed air, are mounted. The nozzle is mounted on a special articulating arm that allows adjusting the flow direction to the current needs for the roadway. The design of mountings makes them possible to adapt to various sizes of the roadway arch support frames, so the spraying

IOP Conf. Series: Materials Science and Engineering **679** (2019) 012019 doi:10.1088/1757-899X/679/1/012019

nozzles can be placed in the mine workings of different cross-section shape. The advantage of the two-media nozzle is the wide fractional range of produced water drops, depending on the parameters of supplied water and compressed air. Such nozzles enable to operate in the range of low water pressure and compressed air pressure (0.05÷0.5 MPa), with low water consumption. The estimated minimum water consumption is 5 dm^3/ h.

4. Principle of operation

The developed solution for the intelligent spraying system, after its installation in the roadway, has to measure PM10 and PM2.5 dust using the EMIDUST ITI EMAG dust meter. Basing on the signals from the dust meter, the pressure parameters for water and compressed air will be selected (one of the nine combinations of settings of water and compressed air parameters), enabling to obtain the highest dust control efficiency below MAC. The hydraulic-pneumatic-electric diagram is shown in Figure 16.

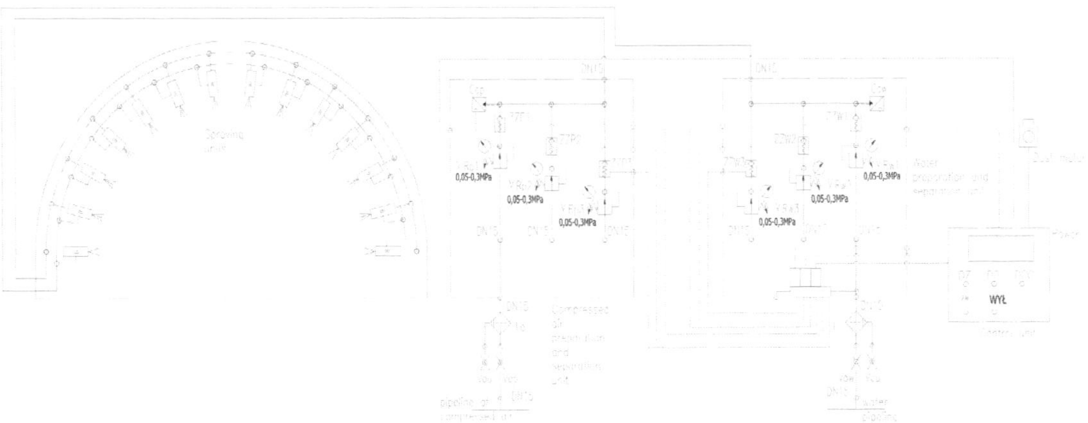

Figure 16. The hydraulic-pneumatic-electric diagram of the intelligent spraying installation.

Water and compressed air will be supplied to water and compressed air preparation and distribution unit from the pipelines. First they will be filtrated and then be distributed into three independent lines controlled by the ZZS check valves. Each water line will have different setting within the range 0.1÷0.3 MPa, and each compressed air line will have different setting within the range 0.2÷0.4 MPa. After getting information from the EMIDUST dust meter, the control unit will select parameters of water and compressed air, and then will open the water and compressed air valves (Figure 17).

IOP Conf. Series: Materials Science and Engineering **679** (2019) 012019 doi:10.1088/1757-899X/679/1/012019

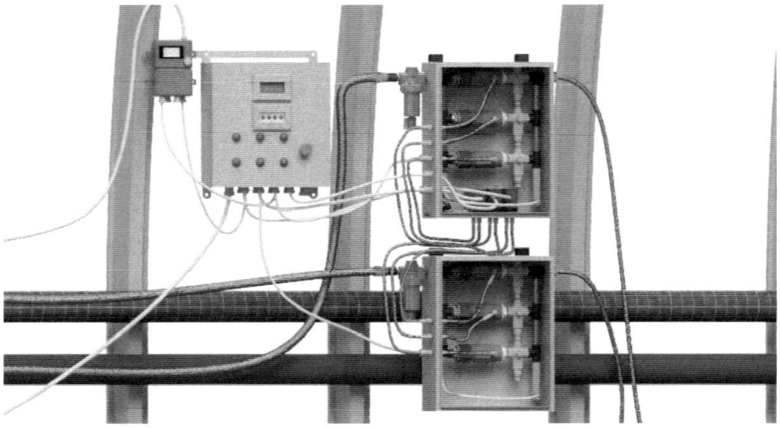

Figure 17. Spraying media control and distribution units.

Proper selection of pressures of water and compressed air supplied to the spraying units will enable the generation of drops with such fractional distributions reduce the identified dust concentration most effectively below MAC.

5. Conclusions

The intelligent spraying installation developed within the ROCD project is a response to the problem of airborne dust, especially the PM10 and PM2.5 fractions, occurring in underground hard coal mines. Its task is to eliminate the unfavourable parameters of spraying installations currently used for dust control, by applying the adaptive actions, consisting in adjusting the operational parameters and drops sizes to the dust concentration measured online. This will allow controlling the amount of water, adjusting it to the actual level of dust concentration. The installation consists of preparation and distribution units for water and compressed air. After passing them the media are distributed to each spraying line of individual settings of water and compressed air. The flow of media to each line of the spraying installation is realized by the controlled check valves. Water and compressed air of the set parameters are directed to the spraying units installed on the support arches, where water mist streams of the most favourable parameters will be generated, with drops size adequate to the measured dust concentration (Figure 18).

Figure 18. Visualization of operation of the intelligent spraying installation.

IOP Conf. Series: Materials Science and Engineering **679** (2019) 012019 doi:10.1088/1757-899X/679/1/012019

Design of the intelligent spraying installation, which will enable the construction of a prototype device will be based on the developed concept. Further stand tests aiming at determination of dust control efficiency depending on the generated water drops parameters [9], which will allow the development of the installation operation algorithm are planned.

References

[1] Bałaga D, Kalita M, Siegmund M and Klimek Z (2019) Measurements of dust concentration work stations of a hard coal mine processing plant after the NEPTUN spraying system. *IOP Conf. Ser.: Mater. Sci. Eng.* **545** 012005

[2] Bałaga D, Kalita M and Siegmund M (2019) Analysis of fraction distribution of the water drops stream generated by the spraying nozzles of new KOMAG design. *IOP Conf. Ser.: Mater. Sci. Eng* . **545** 012010

[3] Ocena stanu bezpieczeństwa pracy, ratownictwa górniczego oraz bezpieczeństwa powszechnego w związku z działalnością górniczo-geologiczną w 2017 roku (2018). Katowice: Wyższy Urząd Górniczy (not published)

[4] Bałaga D, Jedziniak M, Kalita M, Siegmund M and Szkudlarek Z Metody i środki zwalczania zagrożeń pyłowych i metanowych w górnictwie węglowym. *Maszyny Górnicze* 2015 nr **3** pp. 68-81

[5] Frydel W and Steindor M (1996) Instalacje odpylające IO-600 oraz Lutnia Wirowa 700 WIR-700W, Zwalczanie zagrożeń pyłowych w przemyśle górniczym na świecie, Międzynarodowa konferencja naukowo techniczna 17-19 września 1996, Szczyrk

[6] *Rukowodstwo po borbie s pylju w ugolnych szachtach* (1979). Ministerstwo Przemysłu Węglowego ZSRR

[7] Augustyniak Z and Jurkiewicz G (2010) System mgłowy z przysłonami siatkowymi do zwalczania zagrożenia pyłowego". XII konferencja na temat – Problemy bezpieczeństwa i ochrony zdrowia w Polskim górnictwie. Szczyrk

[8] Bałaga D et al. Powietrzno-wodne instalacje zraszające do redukcji zapylenia *Wiadomości Górnicze* 2015 **11** 560-568

[9] Bałaga D and Jaszczuk M 2016 Wpływ parametrów strumienia zraszającego na redukcję zapylenia generowanego przez kombajn ścianowy Prace Naukowe - Monografie KOMAG **47** 1-107

IOP Conf. Series: Materials Science and Engineering **679** (2019) 012020 doi:10.1088/1757-899X/679/1/012020

Utilisation safety of integrated control systems for longwall and road mining complexes.

A Figiel

KOMAG Institute of Mining Technology, Division of Attestation Tests Certifying Body, Pszczyńska 37, 44-101 Gliwice, Poland

E-mail: afigiel@komag.eu

Abstract: Basing on the experience of the certification body in the field of evaluation of integrated control systems for longwall and road mining complexes, as part of the procedure for their authorisation for use in mines, in this article the applicable requirements for these systems have been reviewed (chapter 2), issues important from the safety point of view taken into account by the KOMAG Institute of Mining Technology Division of Attestation Tests Certifying Body in the process of their evaluation have been identified (chapter 3), and measures the fulfilment of which will increase the level of safety in the use of mining and longwall mining complexes in underground mining have been proposed (chapter 4). Issues addressed in this article bear importance for all underground mines, in which the deposit exploitation and excavation is performed using the longwall and road mining complexes.

1. Introduction

Resource extraction in coal mining is performed by longwall mining complexes. Drilling of corridor excavations, to enable exploitation of coal deposits, is done using road mining complexes. Both the aforementioned complexes should be understood as a set of machinery and equipment realizing it's function and ensuring safety of the crew without negative impact on the environment.

The Act on the conformity assessment system [1] defines, among others, the principles of functioning of the system for assessing the conformity of products with the basic and detailed safety requirements. The conformity assessment system consists of regulations defining essential and specific requirements as well as regulations and standards defining activities of entities participating in the conformity assessment process. The conformity assessment process involves manufacturers, their authorised representatives, importers, distributors, certification bodies, inspection bodies and laboratories.

The European Union requirements for the harmonisation of the laws of the Member States relating to machinery intended for use in potentially explosive atmospheres are specified in Directive 2014/34/EU [2], Directive 2006/42/EC [3] and harmonised standards, e.g. PN-EN 60079-25:2011 [4], PN-EN 61508-1:2010 [5].

Division of Attestation Tests, Certifying Body has been involved with the issues of technical safety for several dozens of years, especially with regards to underground mining.

On the basis of many years of experience of the certification body this article points out issues whose inclusion in the design and assessment of integrated control systems for longwall and road mining complexes will contribute to increasing the level of safety.

IOP Conf. Series: Materials Science and Engineering **679** (2019) 012020 doi:10.1088/1757-899X/679/1/012020

2. Requirements

2.1 Definition of longwall and road mining complexes
The mining complex is understood as a set of machinery intended for mining and transportation of coal and securing crew and equipment.

The primary machinery and equipment of the mining complex are:
- powered roof supports control,
- mining machine (longwall shearer or coal plough),
- armoured face conveyor,
- beam stage loader,
- crusher built on a armoured face conveyer or/and beam stage loader,
- pumping unit,
- auxiliary equipment (air conditioners, belt tail pieces of the gate assembly),
- power supply equipment,
- interlock, signalling and loud-speaking communication system,
- security and alarm system equipment.

The primary machinery and equipment of the road mining complex are:
- roadheader,
- belt or chain feeder, installed directly behind the roadheader,
- belt conveyor,
- dust suppression system,
- interlock, signalling and loud-speaking communication system.

Machinery of the longwall and read mining complexes are linked structurally, forming an integrated system. Correct choice of machinery, equipment and systems in a complex (technical potential) is a key factor deciding about the efficiency and productivity of the wall or the efficiency of tunnelling the roadway. It also determines the safety related to utilisation of the longwall mining complex in underground mine workings, which is most important from the view point of market regulatory authorities in the area. Mining complex is built at the destination site – it cannot be tested and assessed before it is transported and assembled on a longwall face and galleries.

2.2 Formal requirements
Particular machines and pieces of equipment of the complex are foremost subject to a conformity assessment with regards to safety requirements specified in separate regulations and verification with regard to compliance with technical specifications required by the user. The assessment of the integrated control system is carried out on the assumption that the individual devices constituting the system meet the requirements for conformity assessment set out in separate regulations [6]. Additionally, with respect to longwall mining and road mining machines, operating in an integrated manner, an assessment of the control system is carried out, which cannot be performed by the manufacturers of individual products (without the knowledge of all machines, equipment and systems forming the mining complex). The assessment of relations between machines that are important for proper operation and safety is made on the basis of technical documentation (project assessment), in most cases developed by a system integrator, which is usually the supplier of power and control equipment acting on behalf of a user (mining plant) or the user himself. The purpose of such assessment, carried out by authorised certification bodies, is to determine whether the control systems of longwall and road mining complexes, when properly installed and used for their intended purpose, do not endanger the safety of people and the environment. On the basis of the results of research and assessment, as well as other formal and legal requirements specified in the regulations, the President of the Higher Mining Authority issues a decision on the approval of the product for use in mining plants [6].

IOP Conf. Series: Materials Science and Engineering **679** (2019) 012020 doi:10.1088/1757-899X/679/1/012020

2.3 Technical requirements

Integrated control systems of the longwall and road mining complexes should conform to technical requirements included in Appendix No. 2, „Technical requirements for products, the application of which in mining plants requires, due to the need to ensure safety of their use in conditions of hazards occurring in mining operations, the issuance of approval" to the Regulation of the Council of Ministers of 30 April 2004 on the approval of products for use in mining plants [7].

The following aspects are taken into account during the assessment of integrated control systems for longwall and road mining complexes:

- provision of information, mainly about:
  - correct operation and technical condition of machinery and equipment,
  - parameters of media influencing operation of machinery and equipment,
  - environmental threats,
  - stopping or blocking of a particular machine and emergency warning,

- ensuring safety in both remote and automatic control modes through:
  - turning on and off of particular machines without creating additional hazards,
  - warning of personnel in the hazardous area with a clear acoustic or optical signal or both,
  - enabling stopping the starting up or stopping and blocking of any piece of machinery,

- the construction of control circuits that:
  - provides protection against electric shock (voltage up to 25 V AC or 60 V DC, SELV or PELV circuits),
  - are resistant to damage which could lead to unintentional starting of the mining equipment or machinery and to blocking the possibility of shutting down the equipment or machinery by the direct control elements and elements controlling the working parameters
  - prevent uncontrolled switching on of the switch as a result of shock and mechanical vibration or stray current influence; in case of voltage decay and it's subsequent return or upon the voltage rising to 1.5 times the nominal voltage
  - work properly when supplied with 0.85 to 1.2 nominal voltage
  - ensure disengagement and blocking of the disengagement state when the resistance of the external loop of the control circuit increases to 600 Ω, the isolation resistance between control wires or between any control wire and earth decreases to 2000 Ω (external circuits) or the resistance of the ground continuity control circuit increases above 100 Ω,

- the machinery stopping device also stops any machinery installed before and after the machinery if its continued operation would create a hazard.

3. Assessment criteria used by the Division of Attestation Tests, Certifying Body

When assessing integrated control systems for mining and face complexes intended for use in mining excavations in potentially explosive atmospheres, in addition to the criteria specified in applicable regulations and briefly discussed in Chapter 2, the Division of Attestation Tests, Certifying Body also performs an assessment of the system with regard to the following factors:

1. the conditions of use of the devices incorporated in the control system
2. selection of machinery and equipment taking into account the level of protection - Group I machinery of category M2 and certified electrical equipment of Group I of category M1 or Group I of category M2 should be used in mining plants
3. selection of electric cables - power supply cables should be chosen taking into account ensuring proper cooperation with the built-in protections in the power supply equipment (overcurrent protection, leakage protection, monitoring of earthing of mobile/replaced machinery)
4. choice of protections in power supply circuits - so that the applied protections work correctly in the system type IT (unearthed system) and protection against electric shock (Earthing system of protective conductors together with continuous monitoring of the insulation status), and so that they meet the requirements of mining standards,

5. compliance with the requirements of mining standards for individual machines (structural and functional requirements) - functions performed by the control system should comply with the functions required by industry standards (European or national)
6. ensuring intrinsically safe electrical system - all intrinsically safe circuits should be assessed in accordance with the requirements of the PN-EN 60079-25:2011 [4] standard. According to this standard, harmonised with Directive 2014/34/EU, intrinsic safety of the system should be presented in a document (document describing the system) in which all electrical equipment, its electrical parameters and the parameters of connecting cables should be specified.

4. Factors contributing to the increase in the level of safety

Many years of experience of the Division of Attestation Tests, Certifying Body, which assesses integrated control systems of longwall and road mining complexes, allow us to formulate problems, the solution of which will contribute to increasing the level of safety of their use.

1. Conformity with the electrical equipment provided by the manufacturer of the machinery.
   Scope of the designer's intervention in the integrated control system of the complex in the control system of a particular machine shall not affect its conformity with the original technical file compiled by the manufacturer. This is a necessary condition for keeping the EU declaration of conformity issued by the manufacturer of the machinery valid.

2. Taking into account the conditions of use of individual machines and equipment in the integrated system.
   This task is only possible if the type of each machine can be identified and its instructions are available. When designing a master system, all conditions, limitations or additional protective measures must be taken into account in order to adapt to the requirements of the machine manufacturer.

3. Powered roof supports control
   Where powered roof supports are operated by hydraulic systems, manually operated roof supports are outside the scope of the integrated system. However, if an electro-hydraulic system is used, whether remotely controlled or in automatic mode, as is the case in a plough wall, then such a system is part of the integrated system and must be assessed in accordance with the applicable regulations.

4. Other equipment for longwall and road mining complexes.
   The integrated control system should include all remote-controlled equipment whose operation is necessary to provide safe operation of the longwall and road mining complex. Additional devices should be assessed not only because of the control mode but also because of mechanical hazards (possibility of reducing the dimensions of the passageway for workers, impact on environmental conditions, etc.).

5. Equipment for control stations.
   Each control station shall be equipped with emergency switch-off facilities by means of which it is possible to switch off the relevant machinery. The emergency shutdown function, as well as other safety functions, shall be performed with a PL performance level or SIL safety integrity level [5]. In particular, it is required that programmed functions are subject to detailed analysis from the functional safety point of view (due to the risk of failure to perform the initiated function).

6. Technological interlocks
   Designing a system, it is easy to introduce technological interlocks, which make the work of a particular machine dependent on others who cooperate with it. An example is a

IOP Conf. Series: Materials Science and Engineering **679** (2019) 012020 doi:10.1088/1757-899X/679/1/012020

technological interlock which makes the operation of a chain scraper conveyor dependent on the operation of a conveyor of a haulage line receiving the excavation, or the operation of a chain scraper conveyor depends on the operation of a armoured face conveyor. Due to the lack of standard requirements, despite the fact that this dependence eliminates the risk of backfilling the receiving conveyor with excavated material, it is not commonly used. In the opinion of the certifying body, the integrated control system should implement all technological interlocks, even if they are not required by standards, and in special, clearly described situations, the operators should be able to deactivate them.

7. Use of control cables in intrinsically safe circuits.
   The documentation describing the intrinsically safe system contains information about the type of control cables and their permissible capacitance and inductance. The user is responsible for checking whether the cables used in a particular control system meet the conditions specified by the designer. A great facilitation for users would be to specify in the documentation the maximum cable length for which the circuit is still intrinsically safe.

8. Control in automatic mode
   Due to potential hazards, automatic mode control should be subject to detailed analysis. The operator should be warned about hazards in time and should have adequate means to stop the dangerous process. For example, the operator of a ladder working near the dump of a longwall conveyor should have at least access to information about the plough head approaching the dump. Despite the lack of such a requirement in mining standards and regulations, technical means implementing the above mentioned function are provided.

5. Summary

The article reviews the current requirements for integrated control systems for longwall and roadhead mining complexes. On the basis of many years of experience in conformity assessment, the Department of Attestation Testing has indicated the aspects the inclusion of which in the process of designing integrated control systems contributes to increasing the level of safety of use of the longwall and roadhead mining complexes. The proposals contained in this article should be taken into account by all entities participating in the process of conformity assessment of each individual mining machines and equipment and in the process of approval of the integrated control system for use in underground mines.

References

[1] Act of 30 August 2002 on the Conformity Assessment System (Journal of Laws of 2019, item 155)
[2] Directive 2014/34/EU of the European Parliament and of the Council of 26 February 2014 on the harmonisation of the laws of the Member States relating to equipment and protective systems intended for use in potentially explosive atmospheres (recast). OJ L 96, 29.3.2014
[3] Directive 2006/42/EC of the European Parliament and of the Council of 17 May 2006 on machinery, and amending Directive 95/16/EC (recast), OJ No L 157, 9 June 2006
[4] PN-EN 60079-25:2011 Explosive atmospheres — Part 25: Intrinsically safe electrical systems
[5] PN-EN 61508-1:2010 Functional safety of electrical/electronic/programmable electronic safety-related systems - Part 1: General requirement
[6] Act of 9 June 2011 - Geological and Mining Law (Journal of Laws of 2019, item 868)
[7] Regulation of the Council of Ministers of 30 April 2004 on the approval of products for use in mining plants (Journal of Laws No. 99, item 1003, as amended)

IOP Conf. Series: Materials Science and Engineering **679** (2019) 012021 doi:10.1088/1757-899X/679/1/012021

Safety of longwall mining with caving in the light of data from monitoring systems

S Rajwa[1]*, Z Lubosik[1] and M Płonka[1]

[1] Central Mining Institute, Plac Gwarków 1, Katowice, Poland

*E-mail: mplonka@gig.eu

Abstract. Shield supports are usually perceived as devices consisting mainly of mechanical and hydraulic parts, while other apparatus, mainly related to electronics, is often treated as a necessary supplement to the system which ensures its control. However, the development of features closely related to the latest electronic equipment in the mining complexes, mainly connected with measuring and registration possibilities, also allows for more and more frequent use of various types of monitoring. It enables observation of the operation of the mining complex in real time or with a slight time delay. What is more, it also enables signalling the operating states of the shield constituting specific threats or irregularities. The most advanced systems found abroad are equipped with functions analysing changes in specific monitored parameters to develop warnings about possible future threats to allow the operator to react with sufficient time in advance. Developing such a forecast system requires gathering and then analysing a sufficiently large and complete database of measurements and observations. These are primarily associated with various difficulties in operating a shield support, and especially with strictly emergency states, such as rock slides, shield clamping conditions, etc. The developed warning identifiers also depend on the mining and geological situation of the specific mining panel. The monitoring currently used, concerning a part of the mining complex related to the shield support, mainly concerns the pressures in the hydraulic props of the shield and the measure of the shifter advance. It allows determining the position of the section on the longwall panel length. The current instantaneous shield geometry and dimensions of the excavation itself, such as height, setting and inclination, are determined in-situ much less often. The information obtained is often also sent to the surface of the mine. The data from monitoring of shield support operation, through proper development of dependencies and analysis results, can optimize the safety of crew, reduce difficulties in maintaining the roof of longwall excavations and thus have a positive impact on the obtained economic results.

1. Introduction

The purpose of shield support, in brief, is its cooperation with the roof as a result of developing adequate bearing capacity and impact on maintaining proper excavation stability. These factors are also dependent on the specific geological and mining conditions characterizing the given mining panel. The selection of a particular type of shield support is usually made earlier by means of analytical calculations. In Poland, for example, the most often used for this purpose is the calculation method based on the theory of deflection of floor layers, developed by A. Biliński [1], which is still being further developed at GIG (determination of roof bearing capacity index 'g').

IOP Conf. Series: Materials Science and Engineering **679** (2019) 012021 doi:10.1088/1757-899X/679/1/012021

Each analytical calculation of the shield support selection always aims to calculate a specific value of the support bearing capacity resulting from its design parameters and the adoption of the excavation dimensions, which has a significant impact on the value of rock mass loading. Until the appearance of effective monitoring systems for shield support operating parameters, it was not possible to verify these assumptions with the conditions of actual operation. Only selected random control measurements were carried out, which were performed sporadically, usually of a modest range during occasional measuring actions. Only when electronic measurements of process parameters could be carried out over a period of time in the range of seconds and concern all or most of the shields, one can speak of effective observation of the operation of the shield support.

Pressure measurements under the piston of hydraulic props of shield support against the time axis are the basic source of information determining the support bearing capacity. In order to determine accurately the bearing capacity of the shield, it is necessary to determine its current geometry. However, due to the high complexity and costs of such measurements during mining, they are very rarely implemented. In order to understand how the shield support works it is necessary to present the operation cycle of the shield.

Most often, measurements of pressure in the sub-piston parts of props and their courses refer to the reference pressure pattern shown in the the Figure1.

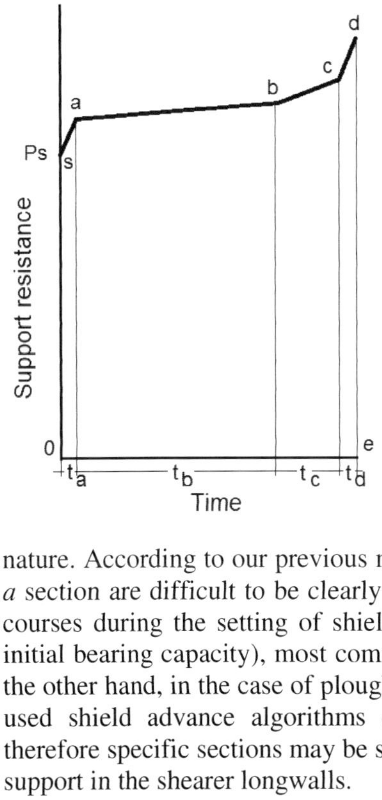

Figure 1. The course of pressure changes in the props during the shield support cycle [2].

Section *0-s* corresponds to the period of support setting, followed by the increase of the bearing capacity in the section *s-a*, caused by the increase in load due to the yielding of the neighbouring shield. Then, a fairly stable and long period of increase in the rock mass pressure in the section *a-b* begins. The section *b-c* shows the influence of the increasing load, as a result of coal cutting to the height of the shield, thereby increasing the extent of the excavation. However, the fast-growing section *c-d* shows the increase in pressure associated with yielding of the neighbouring shield. Section *d-e* shows a sharp drop in the pressure during yielding.

In mining practice, the speed of changes, pressure values, times of particular sections can be subject to large fluctuations, depending on many factors of a geological, mining and technical nature. According to our previous measurement experience, the values for the extreme points of the *s-a* section are difficult to be clearly identified/obtained, due to the very variable shape of the pressure courses during the setting of shield in our mines – generally, due to too low setting pressure (low initial bearing capacity), most common in cases of manual setting of shields in shearer longwalls. On the other hand, in the case of plough longwalls, the shield support is usually run automatically, and the used shield advance algorithms do not always require sequential, subsequent section advance, therefore specific sections may be subject to changes in relation to the presented, model cycle of shield support in the shearer longwalls.

The presented model course can also be changed and supplemented with phenomena such as periods of operation of working valves initiated with pressure drops from the maximum level (subsequent sections of changes similar to the 'saw teeth shape') or with rapid pressure increases due to loads (dynamic phenomena) from the rock mass, initiated by sudden increases in pressure. In cases of individual phenomena, e.g. breaking of rock layers, changes in the corresponding shapes of pressure changes will also be of a one-off, non-periodic nature.

The basic function of the advanced monitoring system is therefore the possibility of using an algorithm for effective identification of all shield operation cycles (based on pressure changes - or e.g.

IOP Conf. Series: Materials Science and Engineering **679** (2019) 012021 doi:10.1088/1757-899X/679/1/012021

hypothetical apparatus signalling of the beginning and end of the cycle) in accordance with the presented diagram of shield support cycle, and possible described changes and modifications that complement it. On the basis of cycle parameters (time and values) determined on its basis, it is only possible to build algorithms for automatic recognition of emergency states and hazards (assuming appropriate limit values for permissible changes). Another issue, after statistical analysis of sets (possibly numerous) of such events, and linking them to various types of obstacles occurring during exploitation, constructing threat prediction system.

Earlier systems, which have less frequent pressure measurements, are less useful for identifying hazards; however, some interesting information can be obtained from them.

2. Example of threat monitoring using systems with smaller frequency of measurements

Longwall A was run at a depth from about 440 m to 530 m, where the seam had a gross thickness of 3.3 to 5.3 m, and its inclination ranged from 15° to 25°. In the roof of the seam there was a layer of slate with coal up to 0.3 m thick, followed by a thin layer of coal as well as claystone and sandy claystone. Above there was a group of strata consisting of mudstone-sandstone layers (7 ÷ 17) m thick.

The floor of the seam was a layer of slate with coal with an underlying claystone. Nevertheless, due to the large variability of the thickness of the seam on the longwall panel length, it was run along coal bottom.

The uniaxial compressive strength (UCS) of both, coal in the seam and the rocks surrounding the seam was characterized by considerable variability.

The compressive strength UCS of the coal seam and surrounding rocks were within the ranges of:

- coal of the seam (7.1 ÷ 10.0) MPa,
- roof of the seam
  - laboratory testing (37.1 ÷ 56.8) MPa,
  - in-situ testing (0.0 ÷ 65.4) MPa,
- floor of the seam (22.8 ÷ 31.4) MPa.

In the analysed longwall panel, disturbances were expected in the seam and surrounding rocks of both sedimentation and tectonic character, causing changes in the thickness of the seam and changes in the stability of the roof rocks. Difficulties in maintaining the roof also occurred locally, which was associated with a deposition of a thin layer of rocks in the direct roof prone to falling after its exposure. It was also found that above the seam numerous mining exploitations were carried out, which also had a significant impact on the roof conditions.

Longall A was run with the longitudinal system with a caving of roof rocks, on the longwall panel length of around 670 m. The width of the longwall was up to 250 m.

In longwall A, the mine used shields with a working height of up to 4.4 m. This shield support was a support and shield structure based on an articulated quadrangle with a lemniscate canopy stabilization system.

The measurement system enabled the pressure measurements to be made with an accuracy of 1 MPa with a frequency of 1 minute in each of the props of all the shields. During the yielding of the shield, measurements were not carried out. Parameters of the system allowed conducting measurements of pressure under the piston of the props in essentially only static conditions. Pressure measurements covered a quantitative range of 160 shields during 208 measurement days.

The measured pressure values under the pistons of the props were referred to the area of the longwall panel as shown in Figure 2. On the chart, a single colourful point with its size corresponds to the widths of the shield over the 1m length of the run. In order to determine the location of the shield, data from the average monthly advance on the mine maps was used.

In places of maximum pressure, maximum loads could be expected from the rock mass during mining operations (which powered support was subjected to), provided that the same working height of the shield and the size of the excavation was maintained. Pressure variation in longwall panel is illustrated in Figures 2 and 3.

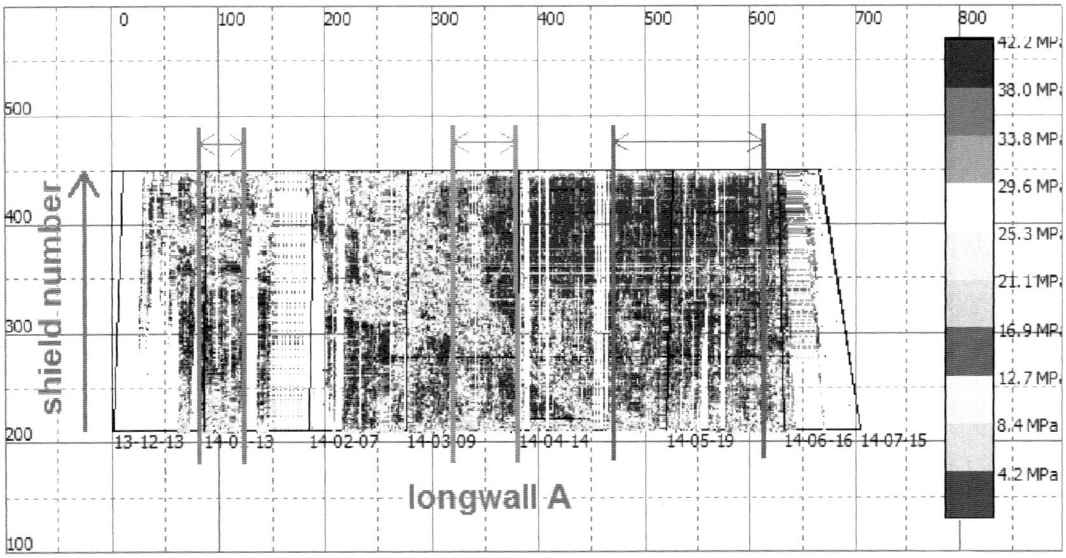

Figure 2. Maximum values of pressures in the area of longwall A with designated fragments of longwall run.

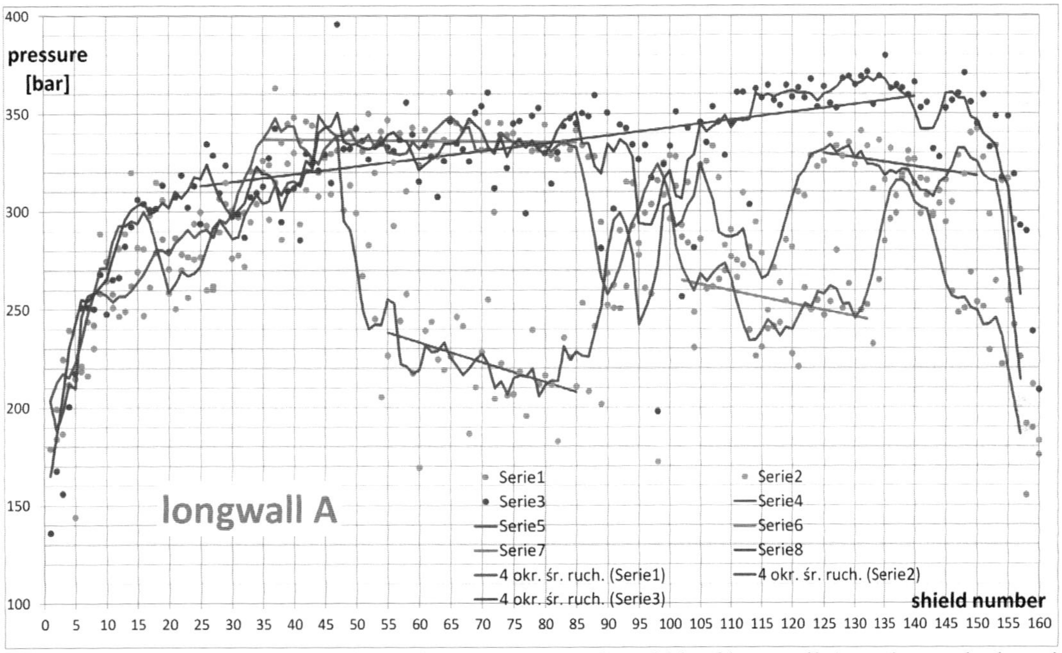

Figure 3. Average maximum values of pressures over the width of longwall A on the marked sections of longwall run.

For three fragments of the longwall A run, where significant variability of pressures at the length of the longwall was observed, marked and limited by lines in brown, green and purple in Figure 2, the average pressure for each shield was calculated and shown in Figure 3 in the form of coloured points. The curves in the same colours were plotted with moving averages computed for four neighbouring shields. However, shorter sections determined from linear regression calculations are additionally marked with fragments of the longwall length characteristic for the pressure levels.

For the 'brown' section of the run, the average variation of the measured pressure levels between section groups 35 ÷ 85 and 102 ÷ 132 reaches about 81 bar (336 bar and 255 bar respectively), while for the 'green' fragment of the run, the average variability of measured pressure levels between groups section 55 ÷ 85 and 125 ÷ 150 reaches about 101 bar (223 bar and 324 bar, respectively). On the purple part of the run, in the section group 25 ÷ 140, the average pressure increases practically uniformly from about 313 bar to 359 bar.

Such an analysis during monitoring may enable determination of areas of variable rock mass loads caused by different floor conditions.

It can further be noted that generally at the longwall ends there are lower pressure values relative to the central part of the longwall. The phenomenon of occurrence of higher loads in the central part of the longwall is represented in many theories and confirmed by means of measurements.

3. Examples of monitoring threats using the identification of measuring cycles in shearer longwalls

Longwall B was run with a roof caving in the seam with a maximum depth of 845 m, with layers inclination from about 16° to 30°. Mainly claystone was located above the seam, locally there were mudstone or sandstone rocks directly in the roof. The floor was a claystone.

According to the obtained data (the longwall was run along coal bottom), the strength of UCS of coal seam and surrounding rocks was:

- coal of the seam 14.9 MPa,
- natural roof of the seam (21.8 ÷ 33.8) MPa,
- floor (coal bottom) 14.9 MPa.

In the exploitation panel of longwall B, in the distance up to 160 m above the seam and below, up to a depth of 60 m, operation has not been carried out so far.

Longwall B was run with a roof caving, with a maximum height (2.8 ÷ 3.0) m, in one-sided vicinity of the goaf. The run of longwall 2 reached about 470 m, and its width was about (190 ÷ 210) m.

Shield support with a working height of up to 4.5 m was used in longwall B.

The measurement system enabled the pressure measurements to be carried out with an accuracy of up to 1 MPa with a frequency of up to 5 seconds in each of the props of all powered support shields. Pressure measurements covered the volume range of 135 shields. The parameters of the measurement system already allow tracking the cycles of the support with a higher frequency (every 5s); however, events that are shorter than 5 seconds may still not be identified.

During the start-up and at the beginning of the longwall run, the courses indicating a decrease in pressure after setting the support, as in Figure 4, there was also the fall of floor layers.

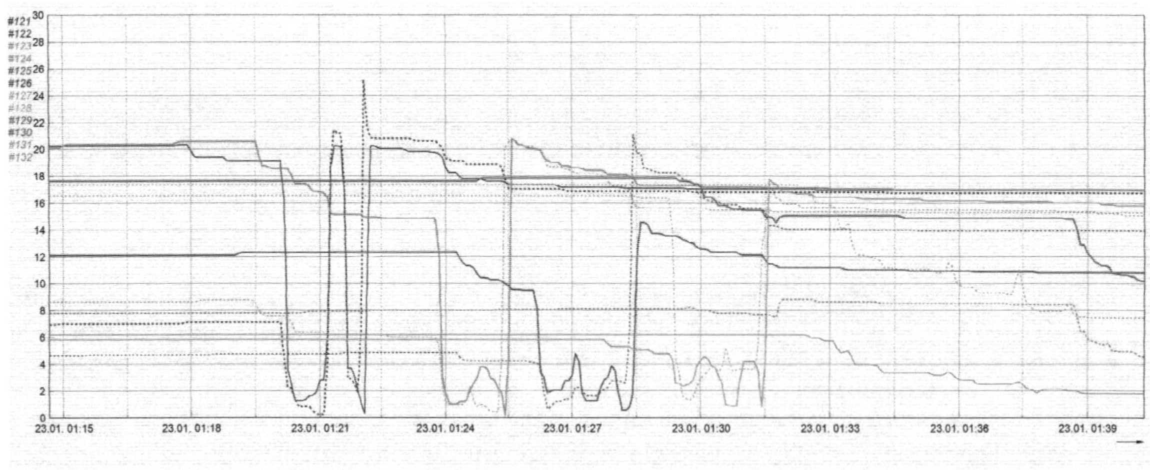

Figure 4. Exemplary courses of decreasing pressure in longwall B (source FAMUR).

A staff congress was organized for the longwall B excavation to assess the situation and further analyses related to pressure courses were conducted. The reasons for the difficulties were complications related to the unfinished longwall start-up and the carelessness of shield setting. On further fragments of the longwall run, and with careful running the interlayer and setting the support, the support cycles obtained the correct form, as in Figure 5.

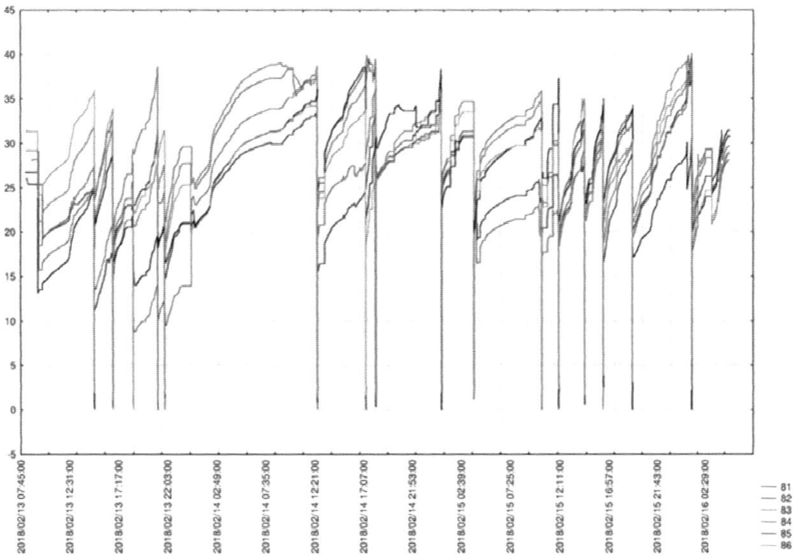

Figure 5. The exemplary course of correct pressure values on longwall B after start-up

During pressure observations in the hydraulic props, it was possible to notice locally operating periods of working valves resulting from floor conditions, as shown in Figure 6, in the form of 'saw-shape' pressure courses before yielding. Another identified phenomenon was the shield groups with increasing or decreasing pressure. The most likely cause of sequential pressure changes were differences in the load of breaking sandstone strata in the direct roof, as in Figure 7.

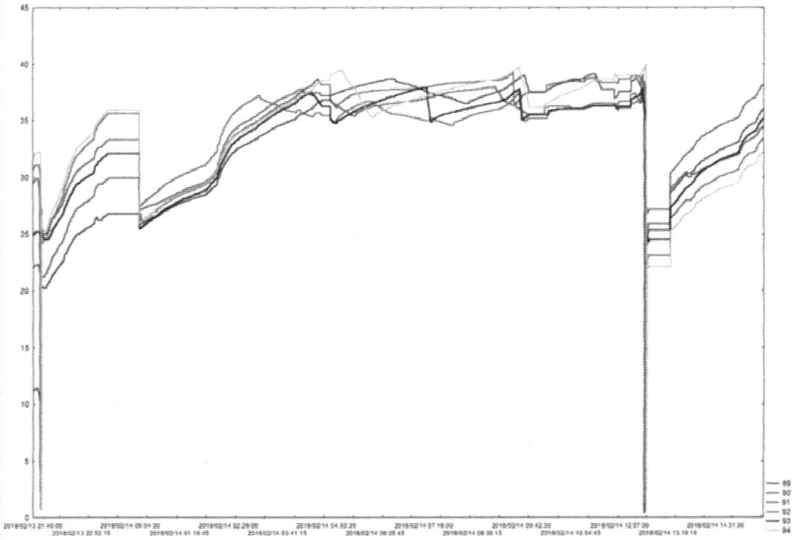

Figure 6. Example pressure waveforms during operation of valves in longwall B.

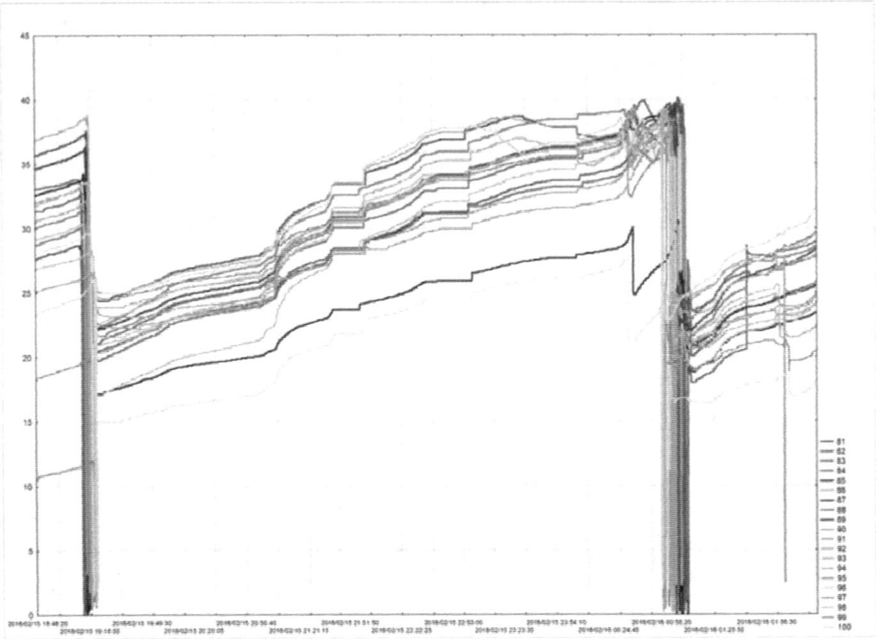

Figure 7. The exemplary course of pressure values presenting the sequence of pressure lowering in longwall B.

4. Examples of threat monitoring using the identification of measuring cycles in plough longwalls

Difficulties associated with the roof maintenance, manifesting themselves in numerous falls of its layers, were noted in the plough longwall C. Due to the similarity of strength characteristics to the longwall D, where the situation was correct and the longwall advance was extremely high, it was decided to compare data from both shield support monitoring systems regarding a single selected measurement day from each mine.

The obstacles in longwall C are illustrated in Figure 8, which indicates the shield numbers at which the roof fall was observed.

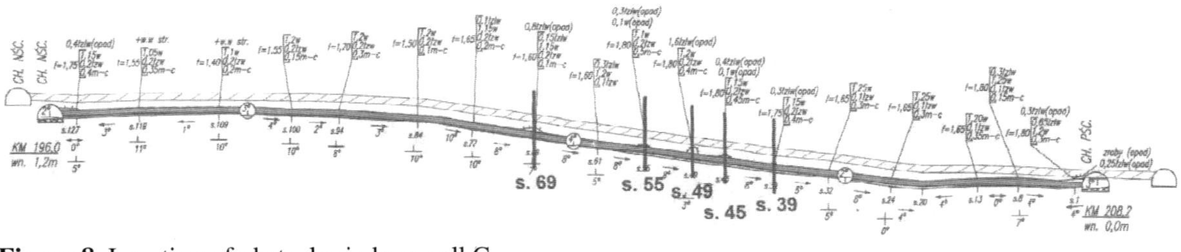

Figure 8. Location of obstacles in longwall C.

Then, the pressure courses for these groups of shields are presented, for one day, in spatial form in Figure 9, where the pressures on the vertical axis are expressed in bars.

IOP Conf. Series: Materials Science and Engineering **679** (2019) 012021 doi:10.1088/1757-899X/679/1/012021

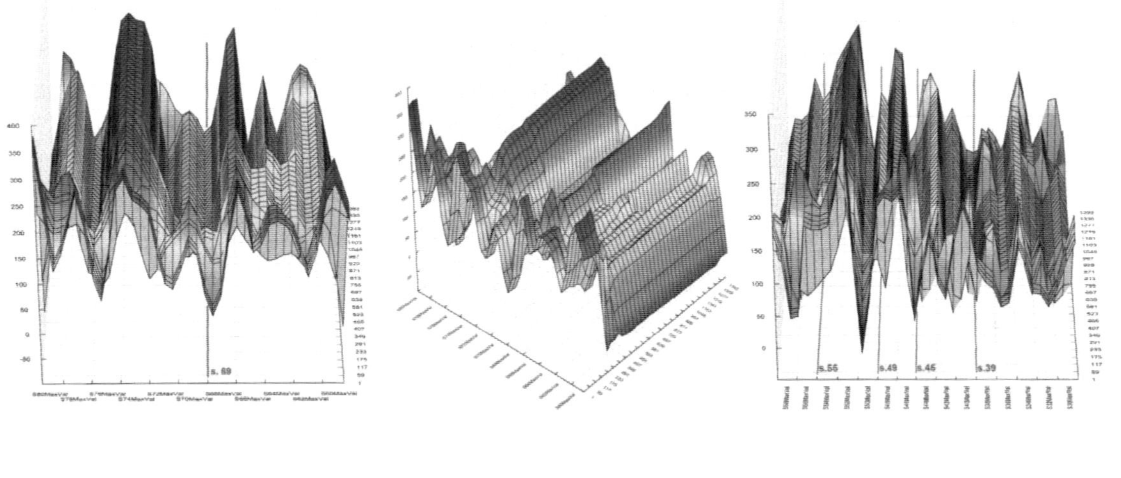

Figure 9. Spatial graph of pressure in longwall C.

Significant differences in the developed pressure can be easily observed between the neighbouring shields. Roof fall was located in areas of reduced pressure. For comparison, spatial charts from longwall D, where the situation was very favourable, are presented in Figure 10.

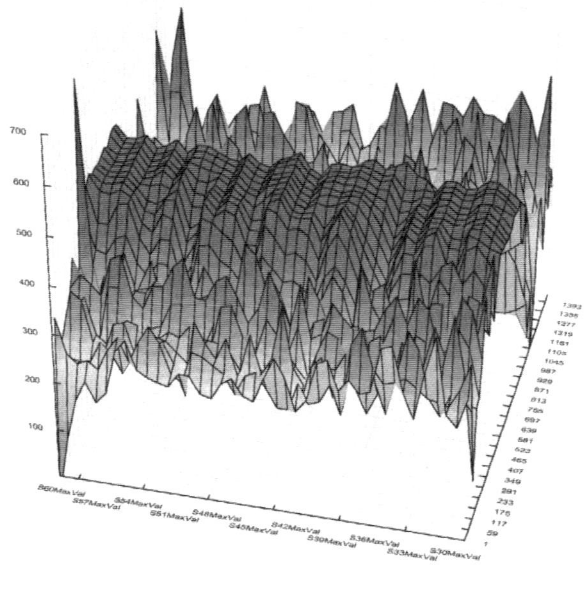

Figure 10. Spatial graph of pressure in longwall D.

Not only a much more even pressure distribution between adjacent sections can be noticed, but also a significantly higher pressure level in general. Pressure differences can be explained to a small extent by slightly different strength conditions and parameters of the types of shield supports used. However, it was difficult to accurately estimate them due to the lack of clearly documented data related to, among others the temporary height of operation in both longwalls. It seems, however, that the main reason for the difficulties in longwall C was the very large diversity of maintained pressures (and thus bearing capacity) between adjacent sections, with their quite low average level calculated along the width of the longwall.

IOP Conf. Series: Materials Science and Engineering **679** (2019) 012021 doi:10.1088/1757-899X/679/1/012021

5. Conclusions

Based on the examples and analyses presented, the following comments can be made:

1. The scope of signalling of possible difficulties is closely related to the technical capabilities and the monitoring software system used.
2. Nevertheless, any properly used monitoring system can provide valuable information on the mining process being conducted.
3. Simple observations of measured pressure levels can easily eliminate basic difficulties related to, e.g. insufficient shield setting at the interlayer or lack of tightness of props.
4. Signalling of operating states with a higher level of complexity requires the determination of a set of limit values for the measured parameters needed to initiate specific warnings. Responsible designation requires the collection of an appropriate measurement and observation database as well as an appropriate analysis to identify them.
5. One can observe a natural desire not to share knowledge about the difficulties occurring during longwall mining. However, in this way we clearly limit the scope of analysis of such states. The question remains whether it possible to achieve a different way of thinking and acting so that knowledge about the possible threats would become the dominant value?
6. Monitoring of shield support and proper data analysis can provide a lot of useful information improving work safety and efficiency of the mining process.

Acknowledgements

The paper uses the results of work carried out as part of the implementation of the statutory work financed by the Ministry of Science and Higher Education: 'Development of a set of methods and calculation indicators for assessing the quality and correctness of the work of powered roof support in the longwall, based on indications of monitoring systems' (Computer symbol of work in GIG No. 11122018 - 152).

References

[1] Biliński A: Metoda doboru obudowy ścianowych wyrobisk wybierkowych i chodnikowych do warunków pola eksploatacyjnego. CMG KOMAG 2005, Gliwice

[2] Peng S S: Longwall Mining. Syd S. Peng publisher, 2006, pp. 183-190

[3] Płonka M eds. (11): Rozkład obciążenia obudowy zmechanizowanej w ścianie zawałowej i jego powiązania z prognozą utrzymania stropu. [Mechanized support load distribution in the caving longwall and its relation to the roof maintenance forecast] Documentation from the implementation of the own research project, GIG documentation No. 04040316-152, unpublished, 2009. Płonka M eds. (11): Charakterystyka podporności obudowy zmechanizowanej w niskiej ścianie kombajnowej w porównaniu do ściany strugowej. [Characteristics of the support of the powered support in the short shearer longwall compared to the plough longwall]

[4] Praca statutowa GIG o symbolu: 11130122-152, 2012, not published

[5] Płonka M, Rajwa S: Assessment of Powered Support Loadings in Plow and Shearer Longwalls in Regard to the Pressure Measurements in Props. International Mining Forum 2011. 24-26 November 2011 – Bogdanka, Poland. A Balkema Book. CRC Press, © 2011 Taylor & Francis Group, London, UK, pp. 221-232

[6] Płonka M, Rajwa S and Lubosik Z Ocena pracy obudowy zmechanizowanej na podstawie danych z systemów monitoringu i postępu sekcji. [Assessment of powered roof support based on data from monitoring systems and section advance] *Przegląd Górniczy* 2017, No. **4**, pp. 25÷33

AUTHOR INDEX